Advanced
Wood
Adhesives
Technology

Advanced Wood Adhesives Technology

A. Pizzi

Ecole National Supérieure des Technologies
des Industries du Bois
Université de Nancy I
Epinal, France
and
University of the Witwatersrand
Johannesburg, South Africa

CRC Press

Taylor & Francis Group

Boca Raton London New York

CRC Press is an imprint of the
Taylor & Francis Group, an **informa** business

CRC Press
Taylor & Francis Group
6000 Broken Sound Parkway NW, Suite 300
Boca Raton, FL 33487-2742

First issued in paperback 2019

© 1994 by Taylor & Francis Group, LLC
CRC Press is an imprint of Taylor & Francis Group, an Informa business

No claim to original U.S. Government works

ISBN-13: 978-0-8247-9266-4 (hbk)
ISBN-13: 978-0-367-40199-3 (pbk)

Library of Congress Cataloging-in-Publication Data

Pizzi, A. (Antonio)
    Advanced wood adhesives technology / A. Pizzi.
        p.   cm.
    Includes bibliographical references and index.
    ISBN 0-8247-9266-1
        1. Adhesives. 2. Wood—Bonding. I. Title.
TP968.P54 1994
668'.3—dc20   94-21042
                                                                                  CIP

Visit the Taylor & Francis Web site at
http://www.taylorandfrancis.com

and the CRC Press Web site at
http://www.crcpress.com

Preface

On my first day as a young, starry-eyed student at the University of Rome, I was profoundly impressed by the sculpted Latin motto on the wall of the great aula in the Faculty of Chemistry:

Poor indeed is the student who does not become better than his teacher.

The problem is that the school taught me about every kind of chemistry, but nothing about adhesives. It is for this reason that this volume has been organized as a textbook, to function as a springboard to the twenty-first century for any wood adhesive technologist in the making. This book is dedicated to the next generation, that they might not need to "reinvent the wheel" as so many of us had to do, in this fascinating and very specialized field. The volume will also be of interest to specialist chemists, engineers, and practicing technologists in this field. It concentrates on adhesives: on the "how" and "why" in adhesives formulation, but at the resin level, not on compounding. Most of the material has not been covered previously. Also, the technology described is such that most of the content must be considered of an advanced nature.

Each chapter begins with an introductory section summarizing the basic aspects of each class of adhesives, to serve as a refresher for experienced scholars in the field, and to set the stage for those who might not have had the benefit of prior study from other books. After a brief, introductory chapter on the nature of adhesion, and adhesion to wood, which is essential for an understanding of this physicochemical phenomenon, the volume is divided into seven additional chapters on wood adhesives classes: UFs, MFs, PFs, RFs, and the newer, emerging adhesives: tannins, lignins, and diisocyanates. Two of the latter, tannins and diisocyanates, have already made an impact and are likely to make even a more profound impact on the future industrial application of wood adhesives. A section presenting some useful, and definitely not elementary industrial adhesive formulations ends each of the last seven chapters. These formulations will be of use to anyone who wants to start in the field above an elementary, inexperienced level.

One of the recurring but not obvious themes throughout the book is the discussion of current or proposed solutions to some of the problems that have bedeviled this field, and that are likely to persist for some time in the future—for example, how to prepare PF resins with pressing times as fast as those of UF resins while maintaining exterior properties, or how to prepare adhesives of low formaldehyde emission or with no formaldehyde or even no aldehydes at all. The focus is never on detailed exposition of the problem: many of my eminent colleagues have already done that. The focus is on feasible, useful, environmentally friendly and ultimately industrially profitable solutions, and on the reasons underlying them. In this the book has a "green" as well as an "industrial" focus on the science and technology of wood adhesives.

On terminating a volume it is always a pleasant duty to thank all those who have participated in it. My first thanks go to the present and earlier generations of wood adhesive technologists, who have contributed over many decades to build the field as it exists today. My thanks are also extended to the staff of Marcel Dekker, Inc., without whom this book would not have seen the light of day.

A. Pizzi

Contents

1
Brief Nonmathematical Review of Adhesion Theories as Applicable to Wood

I. INTRODUCTION

The use of adhesives is a daily occurrence in many wood-processing industries, such as in the particleboard, plywood, and finger-jointing fields. Adhesion is an important physicochemical phenomenon that has attracted considerable attention from many researchers in many fields of science. It is a phenomenon that is noticeable in many materials, not only in relation to what we have come to know of adhesion to a substrate in the narrower sense.

Excellent reviews on adhesion have appeared [1,2], but these tend, first, to generalize principles to all classes of adhesive materials, and as such they communicate little of specific or immediate interest to many applied wood technologists. This is understandable, although unsatisfactory, as adhesives have a wide variety of fields of application in addition to wood. Second, they are definitely written by, and for, adhesion specialists, or at least for people already well versed in the theoretical subtleties of this field. Thus, although excellent, they might not be understood and their message not considered by those who could benefit most by this knowledge: the applied chemists and technologists for whom the industrial use and application of adhesion principles and of wood adhesives is a daily, very real, sometimes very frustrating occurrence. The focus of this chapter is a brief nonmathematical review of the existing theories of wood adhesion so that once understood and appreciated, basic concepts of importance can be built upon by more advanced reading.

II. THEORIES OF ADHESION

In general, four principal theories describing the phenomenon of adhesion are reported:

1.   Mechanical entanglement/interlocking theory
2.   Diffusion theory
3.   Electronic theory
4.   Adsorption/specific adhesion theory

All of these can be used to justify the work that adhesives do in bonding wood. Specifically in the field of wood adhesion, a fifth theory can be added:

5.   The covalent chemical bonding theory

Let's examine these five theories in more detail. First, it is necessary to point out that each theory may be appropriate to a particular class of adhesives under certain circumstances. Often, each mechanism makes a contribution to the adhesion forces at the adhesive–substrate interface. With this in mind, the five theories will be discussed primarily from the perspective of wood adhesion and wood adhesives.

A. Mechanical Entanglement/Interlocking Theory

As the name indicates, this approach proposes that mechanical, physical interlocking of a hardened adhesive into the macro- and microirregularities of a substrate's surface is the major factor in adhesion. However, in many materials good adhesion can be obtained between perfectly smooth surfaces [3–5], indicating that this theory may not be widely applicable. Conversely, there is much information which demonstrates equally convincingly that by increasing the surface roughness of a substrate, apparent increases in bonded joints strengths are recorded. It may well be, however, that mechanical abrasion, such as in veneer sanding or brushing, ensures that substrate surfaces are free of dirt particles, dust, loose fibers, oxidized fibers, and so on. After all, any good adhesive will do a reliable job of bonding the surface that is offered to it, and if this is only a layer of dust, no bonding of the underlying solid surface is likely to occur. Another hypothesis is that roughening the substrate surface increases the interfacial bonding area available, typically by between 5 and 30% and consequently, increases the bonded joint's strength. There is no doubt that all of the above may be important in certain fields of adhesion [6–13]. For instance, adhesion to electroformed copper and nickel sheeting has been proved to be altered by changing the surface topography, with the adhesive's peel strength increasing from 0.66 kJm^{-2} for a totally flat surface to as much as 2.4 kJm^{-2} for a surface formed by 3-μm high-angle pyramids [6]. A correspondence of this effect in wood is not easily found: The reader may think of the need for the use of more adhesive and unusually poorer strengths obtained with very rough veneers in plywood. Equally, changes in the stress distribution at the interfacial region of the joint, induced by substrate roughening, have been suggested as leading to increases in joint strength. This results from preventing microcracks or flaws from propagating along any interfacial line or plane of weakness as rapidly as could be expected in the case of a smooth planar surface.

The bonded joint strength is in practice composed of two parameters: (1) the intrinsic adhesion and (2) the energy, which is viscoelastically and plastically dissipated around the tip of the propagating crack and in the joint's body. This second factor generally dominates the measured joint strength. It is then a case of better stress distribution and dissipation by plastic deformation during joint loading. Evans and Packham [10] and Wang and Vazirani [14] have applied such a concept to an example reminiscent of synthetic thermo-setting and thermoplastic adhesives on a wood substrate. They used a polyethylene/fibrous oxide interface as a composite with discontinuous fibers in a resin matrix. Thus in the case of a wood surface and particularly of a thermoplastic adhesive such as polyvinyl acetate (PVA), when the composite is stressed, stress transfers from resin matrix to wood fiber and back again. This leads to tensile stress in the wood fiber, which is greatest at its center, and to shear stress between the wood fiber and adhesive matrix, which is greatest at the ends. Thus the presence of microfibers on the wood surface would lead to high shear stresses around the ends of the fibers. As a consequence, joint failure occurs by plastic deformation of the cured thermo-plastic adhesive, initially around the wood fiber tips, followed, as the stress concentration is relieved, by failure in the body of the thermoplastic polymer itself. Hence a much larger volume of thermoplastic adhesive will be plastically deformed during fracture than in the case of a completely smooth surface. This higher amount of plastic deformation accounts for the higher joint strength. On beech strips, in tension, PVA can easily reach strengths of 7500 N, whereas resorcinol adhesives, which are much more rigid, reach 4500 N only with difficulty, a clear example of plastic or viscoelastic stress dissipation improving the strength of the joint. In the case of PVA bonding of beech strip surfaces, it is commonly noted that the strength is high but the percentage of wood failure is low. The percentage of wood failure is often limited to shallow, fine fibers and shavings, which could hardly be classified as wood failure. By contrast, in the case of the rigid resorcinol adhesives, notwithstanding the much lower strength needed, there is a high level of deep wood failure.

Furthermore, in the case of wood substrates, and particularly in the case of thermosetting adhesives, a certain amount of adhesive penetration of the first few shallow layers of the wood substrate is desirable. When the adhesive hardens a layer of timber whose empty pores are largely impregnated with hardened adhesives is formed. As is visible under a microscope, a certain amount of mechanical interlocking occurs in the case of timber substrates. There is no doubt that the mechanical interlocking theory might apply to many cases of wood adhesion. Considering what the intrinsic cohesion and brittleness of resins used as thermosetting wood adhesives are when by themselves, however, it is equally certain that although mechanical interlocking might well contribute to joint strength, it does not appear to be the main contributor to wood adhesion.

B. Diffusion Theory

The diffusion theory was promoted in the early 1960s by Voyustskii [15–17]. It states that the intrinsic adhesion of a resin to a polymeric substrate (such as wood) is due to mutual diffusion of polymer molecules across their interface. This suggests that the macromolecules of both the adhesive and substrate, or chain segments of them, possess sufficient mobility and are mutually soluble. This requirement can be expressed by the condition that the polymers of the adhesive and those of the substrate possess similar solubility parameter values, this being a rating of the compatibility of the two materials (e.g., if an amorphous polymer and a solvent have similar solubility parameter values, they should form a solution). The polymer needs to be amorphous, as a high degree of crystallinity tends more to resist dissolving in the solvent (the concept of solubility parameter does not take crystallinity into account).

Although in certain adhesives and substrates there is direct experimental evidence of interdiffusion, several problems exist in applying this theory to wood adhesion. Wood is not a homogeneous substrate; it is primarily a cellular composite of three polymers: cellulose, mostly crystalline but also amorphous, and hemicellulose and lignin, which are both amorphous. It is clear, then, from the solubility parameter concept that some polymers—the amorphous ones, such as hemicelluloses and lignin, and the amorphous portion of cellulose—could undergo mutual diffusion with the polymeric molecules of the synthetic adhesive. The crystalline portion of cellulose is not likely to be involved. This itself indicates that if a diffusion mechanism is involved in wood gluing, it is applicable only to a portion of the wood constituents and to a portion of the substrate's surface. It is also evident that different resins, when in liquid form, such as urea–formaldehyde and phenol–formaldehyde, due to their different chemical structures, are likely to have solubility parameters which are nearer, or more different from, various amorphous components of the substrate. It could be argued that as these resins are primarily in water solution, this may not apply and the relative interdiffusion may be regulated by the relative water-sorption isotherms of the various wood components. This supposition may be slightly far-fetched.

The most fundamental criticism of this theory was raised by Anand et al. [18–23], who suggest that the dependence of the bonded joint's strength on time of contact and resin molecular weight, which are part of the fundamental experimental evidence supporting this theory, can be explained by the effect of wetting of the substrate surface. Anand et al. believe [18–23] that the increase in bonded joint strengths is due to an increasing degree of interfacial contact and that the mechanism of adhesion depends instead on the formation of secondary, van de Waals forces across the interface of adhesive and substrate. In a sense, this is already a rudimentary concept of one of the theories that follows, the adsorption theory. In conclusion, where polymers are highly cross-linked, which is the case with thermosetting wood adhesives, and where they are highly crystalline, such as in crystalline wood cellulose, interdiffusion is an unlikely mechanism of adhesion.

When an elastomer (such as PVA) interfaces with an amorphous polymeric substrate constituent, if the solubility parameters of the two substances are similar, a certain amount of interdiffusion may be possible, although its contribution to total adhesion is likely to be low. There is, however, one case of wood bonding in which interdiffusion appears to exist and is likely to play a measurable role—in the production of fiberboard that does not contain synthetic adhesive. Under high levels of moisture content, high pressure, and a long pressing time, the glass transition temperature of lignin is lowered, the lignin in the fibers is mobilized, and interdiffusion between lignin polymers or polymer segments on different fibers contributes to bonding the fiberboard together. Interdiffusion does appear to play a quantifiable role in this case. It is equally clear that according to adsorption theory exposed later, secondary forces still appear to be the primary contributors to adhesion.

C. Electronic Theory

The electronic theory was promoted mainly by Deryaguin et al. [24–26], who propose that if adhesive and substrate have different electronic band structures, there is likely to be electron transfer upon contact of the two surfaces. This results in the promotion of a double layer of electrical charges at the interface of the adhesive with the substrate. They suggest that the electrostatic forces arising from this contact of surfaces may contribute significantly to adhesion. The latter statement has caused considerable controversy [27–30], as it implies that these electrostatic forces are an important cause rather than just a result of high joint strengths. This theory in practice treats the adhesive–substrate coupling as a capacitor which is charged due to the contact of two different materials. Separation of the two surfaces of this capacitor leads to a separation of charges that becomes increasingly severe, until a discharge occurs. Adhesion is presumed to be the consequence of the existence of these attractive forces across the electrical double layer.

Criticism of this theory is quite extensive. While the proposers of this theory suggest that an electron transfer mechanism from substrate to polymer is at play and is responsible for intrinsic electronic adhesion, critics suggest that this should be a fairly rapid process, and althought it might even explain initial adhesion, it cannot explain the slow buildup in adhesion that is observed during aging. Although surface electronic charges are often not found on substrate surfaces, electron emission is altered, indicating that the surface electronic states of substrate and adhesive are altered [31]. For instance, surface charge densities are apparent when rubber is separated from a substrate [32], and these charges are enough to attract dust, resulting in reduced adhesion if recontact is attempted. However, the contribution of these charges to total adhesion was found to range from negligible to considerably less than 10%. In Roberts' experiments [32], for instance, the contribution from the electrical double layer at a rubber–glass interface was found to be about 10^{-5} mJm^{-2}, which is truly negligible compared to the contribution of, for instance, van der Waals forces, which is 60 mJm^{-2}.

Application of this theory to wood adhesion does not appear possible, and no experimental evidence of the existence of this contributory factor has ever been recorded for wood adhesion. It appears that for typical adhesive–substrate interfaces, any electrical double layer that might be generated does not contribute significantly to intrinsic adhesion. Any electrical phenomenon observed during the joint fracture process probably arises from the bond rupture itself rather than being the cause of adhesion between the two materials.

D. Adsorption/Specific Adhesion Theory

In the wood adhesion field, the adsorption theory of adhesion, sometimes called the theory of specific adhesion, states that an adhesive will adhere to a substrate because of the intermolecular and interatomic forces between the atoms and molecules of the two materials. This theory is definitely the most widely accepted and applicable theory of adhesion. In the theory's widest sense, the intermolecular and interatomic forces between adhesive and substrate can be of any type. Thus secondary forces such as van der Waals, hydrogen bonds, and electrostatic forces are equally acceptable as ionic, covalent, and metallic coordination bonds. In wood adhesion, however, the age-old and mistaken belief that in wood bonding covalent linkages must be present to ensure good joint strength has caused this theory to be split into an adsorption/specific adhesion theory proper, in which only the effect of secondary forces is taken into consideration, and in a covalent chemical bonding theory in which good adhesion is ascribed only to the presence of covalent bonding between adhesive and substrate. This separation is, of course, fictitious and is unfortunately at variance with what is accepted in all other fields of adhesion. As this review is aimed particularly at wood adhesion, these two different aspects of the same theory are discussed separately. In this section only the influence of secondary forces onto wood adhesion will be reviewed. First, it is necessary to indicate the range of bond energies characteristic of secondary and primary forces. These are indicated in Table 1.1 [1,33–35].

Several authors [36–39] have calculated the attractive forces between two planar bulk phases due solely to secondary forces. Their results show that such attractive forces would result in a joint strength in tension of over 100 MPa. This is considerably higher than the experimental strength of most adhesive joints. The discrepancy can be ascribed to the presence of air-filled voids, defects, or geometrical features causing fracture of the joint at stresses very much lower than the value calculated theoretically. However, these calculations clearly indicate that high joint strengths can be achieved, in theory, from the adhesion that results exclusively from secondary forces acting across the adhesive–substrate interface. The theoretical conclusion that a multitude of secondary forces is then sufficient to account for the average experimental joint strengths attained is supported by a considerable body of experimental results [40–56] demonstrating that the principal mechanism of adhesion in many adhesives and substrates involves only interfacial secondary forces. In the case of wood adhesion, all three principal types of secondary forces—van der Waals, hydrogen bonds, and electrostatic interactions—appear to play a role.

Table 1.1 Bond Types and Typical Bond Energies

Type	Bond energy $(kJ\ mol^{-1})$
Primary bonds	
Ionic	600–1100
Covalent	60–700
Metallic, coordination	110–350
Donor–acceptor bonds	
Brønsted acid–base interactions	Up to 1000
(i.e., up to a primary ionic bond)	
Lewis acid–base interactions	Up to 80
Secondary bonds	
Hydrogen bonds (excluding fluorine)	1–25
Van der Waals bonds	
Permanent dipole–dipole interactions	4–20
Dipole-induced dipole interactions	Less than 2
Dispersion (London) forces	0.08–40

Source: Data from Refs. 1 and 33 to 35.

Recently, the adhesion of both phenol–formaldehyde (PF) resins to crystalline wood cellulose and of urea–formaldehyde (UF) resins to both crystalline and amorphous wood cellulose have been modeled as a physicochemical adsorption due entirely to the balance of attractive and repulsive secondary forces at the interface between adhesive and substrate [57–61]. In these studies the theoretical adhesions forces of different PF and UF oligomers to cellulose substrates were calculated. It was found that adhesion of these resins to a cellulose substrate can be enhanced by shifting the conditions of preparation of the resin to favor the proportion of species having higher specific adhesion. A variety of other interesting conclusions corresponding to existing experimental evidence were also shown from these investigations [57–61]. The correspondence of these theoretical calculated results with experimental results appears to confirm that adsorption/specific adhesion by secondary forces is indeed the most relevant mechanism operating in wood adhesion. From these theoretical results a method for the approximate quantification of total adhesion of UF resins to lignocellulosic substrates was devised [60]. The method allowed simplified screening of UF formulations to optimize their adhesion. Its application to known practices in UF resin preparation also indicated good correspondence with direct experimental evidence. The same can be said for UF adhesives [58,59].

There is no doubt that the van der Waals forces' contribution is the more pronounced one for most PF dimers and methylol phenols on crystalline wood cellulose, as can be seen from Table 1.2. However, the contribution of hydrogen bonding is also considerable in some cases: in Table 1.2 the relative contributions of van der Waals forces and hydrogen bonding in the case of PF resins to crystalline cellulose are indicated. There is, furthermore, a considerable body of experimental literature detailing and demonstrating the importance of hydrogen bonding. Of particular relevance is the work of Kusaka and Suetaka [49], who have employed reflectance infrared (IR) spectroscopy to study the interfacial bonding between a cyanoacrylate adhesive and anodized aluminum. They observed a lowering of the C–O stretching frequency and a shift in the asymmetric stretching vibration of the C–O–C group to a higher frequency in the cyanoacrylate infrared spectrum once adsorbed on the aluminum surface. These changes were interpreted as being due to the formation of interfacial hydrogen bonds between the carbonyl groups on the cyanoacrylate and the hydroxy groups on the aluminum oxide surface. There are many other cases, with totally different adhesives and substrates, in which the importance of hydrogen bonding has been demonstrated experimentally [50–56]. Mechanisms of adhesion by hydrogen bonding can include direct hydrogen bonding between adhesive and substrate groups, as well as hydrogen bonding of an adhesive group to a water molecule which is also hydrogen bonded to a chemical group of the substrate [52]. Both mechanisms apply in wood adhesion, where even after the adhesive has hardened and the joint has been humidity conditioned, $\pm 10\%$ water remains in the wood. Supporting evidence for the existence of a variety of hydrogen-bonding mechanisms comes from Blythe et al. [56] using x-ray photoelectron spectroscopy. In the case of wood adhesion the contribution of the electrostatic component is, instead, rather small, and this has the tendency to be repulsive rather than attractive, as can be seen from Table 1.2. The dielectric constant of water has been taken into account in these calculations, as it should be, due to the residual moisture content in the wood substrate. If it is not, the electrostatic forces present extremely high values, leading to absurd results.

Donor–acceptor interactions (Table 1.1), defined as the formation of acid–base interactions between adhesive and substrate, have also been proposed [62–67] as a major type of intrinsic adhesion force operating across the interface. The proposers of this concept, however, include hydrogen bonds as a subset of acid–base interactions. The acid may be a Lewis electron donor or a Brønsted proton acceptor. Even equations describing the work of adhesion for

Table 1.2 Relative Contributions (kcal mol^{-1}) of Van der Waals, Hydrogen Bonds, and Electrostatic Secondary Forces to the Adhesion of Species from a PF Resin to Wood Cellulose[a]

	Van der Waals	Hydrogen bonds	Electrostatic	Total energy of cellulose–phenol interaction
o-Monomethylol phenol	−12.60	−0.23	0.64	−12.190
p-Monomethylol phenol	−10.77	−0.46	0.66	−10.567
o,o-Dimethylol phenol	−10.45	−0.77	0.67	−10.550
o,p-Dimethylol phenol	−12.47	−10.00	0.65	+3.119
PF dimers				
o,o-Dimer	−5.34	−7.75	0.77	−12.329
o,p-Dimer	−7.00	−5.81	0.73	−12.080
p,p-Dimer	−6.51	−4.37	0.68	−10.203

[a]Negative values indicate attractive secondary forces; positive values indicate repulsive secondary forces.

donor–acceptor interactions have been reported [63]. Unfortunately, the same authors could not establish whether increases in adhesion arise from acid–base rather than dipole–dipole interactions, as the predictive capability of the proposed equations is very restricted due to lack of critical data. Other authors' work [62,68], furthermore, indicates that even when the values of these critical data are available, the equations proposed are incapable of predicting acid–base interactions at the interface; instead, they predict that dipolar interactions are not negligible, as theorized by the equation's proposers.

E. Covalent Chemical Bonding Theory

As already explained, the covalent chemical bonding theory does not really exist in any field of adhesion other than wood adhesion and can be considered as the other main subset of adsorption theory. There is in the literature on wood adhesion [69–77] a belief that good joint strengths can be justified exclusively if covalent bonds are present between adhesive and substrate. Johns has reviewed this belief in depth [78] but exclusively from the point of view of the existence, or not, of strictly covalent chemical bonding. The general adsorption theory, instead, defines adhesion by primary bonds as possible through not only covalent but also ionic and metallic coordination bonds (Table 1.1). It will then be necessary to deal, first, with the concept of strictly covalent chemical bonding between adhesive and substrate, and with the evidence in favor and against it, and then with the contribution of ionic and metallic coordination bonds, which are important in the more generalized adsorption theory and also in wood adhesion. Strictly covalent bonds appear to occur in certain fields of adhesion, but most of the evidence presented suggests that to have adhesion by establishing strong primary bonds across the interface requires the use of special techniques, such as the incorporation of particular reactive groups in the adhesive, the substrate, or both. The concept that is introduced, and the necessary condition which then needs to be satisfied, is that the adhesive and substrate must be capable of reacting chemically with each other; thus such type of adhesion is applicable only to mutually reactive materials: not a common characteristic in wood adhesion. For example, infrared evidence has been found of covalent primary bonds between a polyurethane adhesive and a reactive epoxy-based primer [79] and between cross-linking elastomers on reactive substrates [40,41,80]. In all fairness it should be noted that in these cases the covalent bonds formed have a high degree of ionic character, due to the different electronegativities of the atoms involved in the bond. Interfacial primary bonds which are highly ionic in character have also been reported (such as the $-COO^-/A^+$ type) [81,83].

In his excellent review on covalent chemical bonding in wood adhesion, Johns [78] arrives at the conclusion that such types of bonds do not exist between adhesive and wood substrates. The classical example of what was thought to give covalent chemical bonding are diisocyanate adhesives such as diphenylmethane-4,4'-diisocyanate (MDI). Here, direct reaction of the isocyanate group with the several types of hydroxy groups present on lignin and particularly on wood carbohydrates was suggested [72,74,78] but never demonstrated. When one notes that the reaction of polymeric MDI, a diisocyanate adhesive, with water [84] to form polyureas proceeds at 7.4×10^{-6}L mol^{-1} s^{-1}, while the rate constant of its reaction with wood carbohydrate alcoholic and aliphatic hydroxy groups [84] such as in lignin are 2×10^{-7} and 6×10^{-6}L mol^{-1} s^{-1}, respectively, it is evident that covalent chemical bonds between an isocyanate group and wood constituents are very unlikely under normal application conditions for wood adhesives. The isocyanate will react so much faster with water, and wood always contains residual moisture content, to partly deny such covalent bonding possibility. MDI with water was shown, instead, to form polyureas [85], very strong adhesives, adhering to the substrate by secondary forces alone.

Urea–formaldehyde, melamine–formaldehyde, and phenol–formaldehyde resins have also been reputed, at some stage, to present covalent chemical bonds attained by reaction of their active methylol groups with the hydroxy groups of carbohydrates or with the aromatic nuclei of lignin [70,71]. As Johns [78] remarks, the demonstrated reaction of aldehyde resins and wood material is not surprising: wood is a chemically reactive system capable of reacting with another reactive system, the adhesive, when placed in a high-temperature environment. Under such, in a sense extreme, conditions, many reactioins of the methylol group of the adhesive with wood constituents can be thought of. However, the problem becomes one of determining if covalent bonds between adhesive and substrate exist, and to what extent, under normal wood adhesive operating conditions. Under normal conditions the formation of covalent bonding between adhesive and wood substrate has never been observed [78]. This does not mean that it does not exist: it only means that either covalent bonds are not present or they are in such a low proportion as to be undetectable. Hence their contribution to wood adhesion is absent or, at best, negligible.

There are many other arguments against the formation of proper covalent bonds between adhesive and wood substrate, but it may be of interest here to illustrate only one of them, involving UF resins. Figure 1.1 illustrates gel time as a function of the pH values of phenolic and UF resins. Admitting the UF to be the adhesive and the phenolic resin to be the phenolic moieties of lignin, it is easy to see from the figure that covalent reaction of the methylol group of the UF resin with the phenolic nuclei of lignin is impossible under

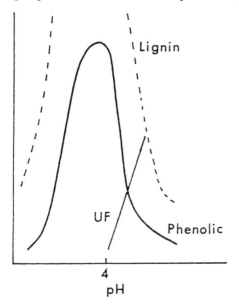

Figure 1.1   Gel time as a function of pH of UF and PF resins, and of lignin with formaldehyde.

normal wood adhesive operating conditions: at lower pH values, self-poly-merization of the UF adhesive is too fast, and thus coreaction of its methylol groups with the phenolic moieties of lignin is impossible (or of very low probability). At higher pH values the UF resin does not harden, and thus even if coreaction is possible, the resin does not function as an adhesive. Only in the narrow band of pH values around the crossover point of the two gel-time curves might coreaction occur, but who can afford the very long gel times needed to cure the UF resin at pH values of around 7? It is then clear that although potentially, under very particular reaction conditions, covalent bonds between UF adhesives and wood can occur, under normal wood adhesive reaction conditions they do not, or if they do, their contribution is truly negligible. The conclusion is identical for PF adhesives, the methylol groups of which reacts much faster with the aromatic nuclei of the adhesive than with the aromatic nuclei of lignin.

All of the discussion above deals strictly with covalent bonds. There are, however, clear and demonstrated cases of primary bonds, other than covalent ones, in the case of wood adhesion. This is the case of the metallic coordination bond, which is central to the promotion of molecular complexes. There is no doubt that these exist in fields other than wood adhesion and there is abundant experimental proof for them [1]. It is difficult to define the term *molecular complex* precisely, but in short it is applied to cases in which the number of bonds formed by one of the atoms is greater than expected from its usual valence considerations. The atoms' groupings, or ions, which bond to the central metallic atoms are referred to as *ligands* and are characterized according to how many electron-donor centers they possess. Ligands can bond through only one atom (e.g., N in NH_3), or two or more on the same molecule. Consider now the case of wood treated with the wood preservative CCA (copper–chromium–arsenate). Copper easily forms stable molecular com-plexes with the *o*-diphenol and *o*-methoxyphenol moieties of lignin [86,87]. Chrome in both its trivalent and hexavalent states forms similar complexes [88–90, 91–94], and in its CrO^{4-} form is able to react covalently with both the carbon–carbon double bonds and the aromatic nuclei of lignin [88,91–94]. Due to the bidentate structure of the ligand, these *o*-diphenol complexes are quite stable.

It is clear that due to the lignin ligand moieties being placed on a fixed network (the wood matrix), each Cu^{2+} atom is in principle able to coordinate two phenolic or methoxyphenolic moieties of lignin, but in reality, due to their relative distances, only one aromatic nuclei of lignin will attach to one Cu^{2+}. Thus Cu^{2+} then has the ability to coordinate any other mobile ligand that can reach it. If the adhesive, in its liquid state, does contain electron-donor groups such as amidic (UF resins) or phenolic groups (PF resins), it will do just this. On any surface where the copper atoms are complexed to lignin, the adhesive mobile phase will also complex to the copper by coordination bonds.

Approximately the same is valid for chrome. As normal PF and resorcinol–formaldehyde resins are only capable of forming monodentate complexes, these bonds will not be particularly strong [97]. Where, instead, the adhesive presents *o*-diphenol structures such as in tannin adhesives [95–97], bidentate structures are formed and these bonds are very strong. It is interesting to note that this situation can lead to stronger joints or to very much poorer joints than in untreated timber.

III. CONCLUSIONS

It is clear that under particular circumstances any of the primary mechanisms of adhesion may be responsible for a bonded joint's strength. In the case of wood adhesion, specific adhesion by secondary forces always appears to be the predominant component. Mechanical interlocking is also likely to be present in most cases, but its contribution to total wood adhesion is definitely much lower. In particular cases adhesive–substrate interdiffusion might even contribute to wood adhesion, although for most normal wood adhesives its contribution appears to be negligible. Metal coordination bonding can contribute, at a low level, to the bonding of timber surface treated with waterborne inorganic preservatives such as CCA. Instead, strictly covalent bonding appears to be negligible or, probably, absent completely. Too little has been done in the wood adhesion field to evaluate the contribution of the electronic theory. However, if what is found in other adhesive fields can be transferred to wood adhesion, its contribution also appears to be so negligible that this theory might not be applicable at all to this particular field.

REFERENCES

1.   A. J. Kinloch, *Adhesion and Adhesives Science and Technology*, Chapman & Hall, London, 1987.
2.   W. C. Wake, *Adhesion and the Formulation of Adhesives*, 2nd ed., Applied Science Publishers, New York, 1982.
3.   D. Tabor and R. H. S. Winterton, *Proc. Roy. Soc.*, *A312*: 435 (1969).
4.   J. N. Isrealachvili and D. Tabor, *Proc. Roy. Soc.*, *A331*: 19 (1972).
5.   K. L. Johnson, K. Kendall, and A. D. Roberts, *Proc. Roy. Soc.*, *A324*: 301 (1971).
6.   D. J. Arrowsmith, *Trans. Inst. Met. Finish.*, *48*: 88 (1970).
7.   J. R. Evans and D. E. Packham, in *Adhesion*, Vol. 1 (K. W. Allen, ed.), Applied Science Publishers, London, 1977.
8.   J. R. Evans and D. E. Packham, *J. Adhes.*, *9*: 267 (1978).
9.   J. R. Evans and D. E. Packham, *J. Adhes.*, *10*: 39 (1979).
10. J. R. Evans and D. E. Packham, *J. Adhes.*, *10*: 177 (1979).
11. D. E. Packham, in *Developments in Adhesives*, Vol. 2 (A. J. Kinloch, ed.), Applied Science Publishers, London, 1981.
12. D. E. Packham, in *Adhesion Aspects of Polymeric Coatings* (K. L. Mittal, ed.), Plenum Press, New York, 1983.
13. P. J. Hine, S. El Muddarris, and D. E. Packham, *J. Adhes.*, *17*: 207 (1984).
14. T. T. Wang and H. N. Vazirani, *J. Adhes.*, *4*: 353 (1972).
15. S. S. Voyutskii, *Autohesion and Adhesion of High Polymers*, Wiley-Interscience, New York, 1963.
16. S. S. Voyutskii, *Adhes. Age*, *5*(4): 30 (1962).
17. S. S. Voyutskii, Yu I. Markin, V. M. Gorchakova, and V. E. Gul, *Adhes. Age*, *8*(11): 24 (1965).
18. J. N. Anand and M. J. Karam, *J. Adhes.*, *1*: 16 (1969).
19. J. N. Anand and R. Z. Balwinski, *J. Adhes.*, *1*: 24 (1969).
20. J. N. Anand, *J. Adhes.*, *1*: 31 (1969).
21. J. N. Anand and L. Dipzinski, *J. Adhes.*, *2*: 16 (1970).
22. J. N. Anand, *J. Adhes.*, *2*: 23 (1970).
23. J. N. Anand, *J. Adhes.*, *5*: 265 (1973).
24. B. V. Deryaguin, *Research*, *8*: 70 (1955).
25. B. V. Deryaguin, N. A. Krotova, V. V. Karassev, Y. M. Kirillova, and I. N. Aleinikova, *Proceedings of the 2nd International Congress on Surface Activity*, Vol. III, Butterworth, London, 1957.
26. B. V. Deryaguin and V. P. Smilga, in *Adhesion: Fundamentals and Practice*, McLaren and Son, London, 1969.
27. C. Weaver, in *Adhesion: Fundamentals and Practice*, McLaren and Son, London, 1969.
28. C. Weaver, in *Aspects of Adhesion*, Vol. 5 (D. J. Alner, ed.), University of London Press, London, 1969.
29. C. Weaver, *Faraday Spec. Discuss.*, *2*: 18 (1972).

30. C. Weaver, *J. Vac. Sci. Technol.*, *12*: 18 (1975).
31. C. T. H. Stoddart, D. R. Clarke, and A. J. Kirkham, *J. Adhes.*, *2*: 241 (1970).
32. A. D. Roberts, in *Adhesion*, Vol. 1 (K. W. Allen, ed.), Applied Science Publishers, London, 1977.
33. L. Pauling, *The Nature of the Chemical Bond*, Cornell University Press, Ithaca, N.Y., 1960.
34. R. J. Good, in *Treatise on Adhesion and Adhesives*, Vol. 1 (R. L. Patrick, ed.), Marcell Dekker, New York, 1967.
35. F. M. Fowkes, in *Physicochemical Aspects of Polymer Surfaces*, Vol. 2 (K. L. Mittal, ed.), Plenum Press, New York, 1983.
36. J. R. Huntsberger, in *Treatise on Adhesion and Adhesives*, Vol. 1 (R. L. Patrick, ed.), Marcel Dekker, New York, 1967.
37. J. R. Huntsberger, *Adhes. Age*, *13*(11): 43 (1970).
38. E. Orowan, *J. Franklin Inst.*, *290*: 493 (1970).
39. D. Tabor, *Rep. Prog. Appl. Chem.*, *36*: 621 (1951).
40. E. H. Andrews and A. J. Kinloch, *Proc. Roy. Soc.*, *A332*: 385 (1973).
41. E. H. Andrews and A. J. Kinloch, *Proc. Roy. Soc.*, *A332*: 401 (1973).
42. A. N. Gent and A. J. Kinloch, *J. Polym. Sci.*, *A2*: 659 (1971).
43. J. P. Bell and W. T. McCarvill, *J. Appl. Polym. Sci.*, *18*: 2243 (1974).
44. A. F. Yaniv, I. E. Klein, J. Sharon, and H. Doduik, *Surf. Interface Anal.*, *5*: 93 (1983).
45. R. A. Gledhill and A. J. Kinlock, *J. Adhes.*, *6*: 315 (1974).
46. A. J. Kinloch, *J. Adhes.*, *10*: 193 (1979).
47. D. K. Owens, *J. Appl. Polym. Sci.*, *18*: 1869 (1974).
48. W. E. Baldwin, A. J. Milun, and H. A. Wittcoff, *Prepr. Am. Chem. Soc., Div. Org. Coat. Plast. Chem.*, *29*: 30 (1969).
49. I. Kusaka and W. Suetaka, *Spectrochim. Acta*, *36A*, 647 (1980).
50. T. Furukawa, N. K. Eib, K. L. Mittal, and H. R. Anderseron, *Surf. Interface Anal.*, *4*: 240 (1984).
51. W. H. Pritchard, in *Aspects of Adhesion*, Vol. 6 (D. J. Alner, ed.), University of London Press, London, 1970.
52. A. Baskin and L. Ter-Minassian-Saraga, *Polymer*, *19*: 1083 (1978).
53. B. Leclerq, M. Sotton, A. Baszkin, and L. Ter-Minassian-Saraga, *Polymer*, *18*: 675 (1977).
54. D. K. Owens, *J. Appl. Polym. Sci.*, *19*: 265 (1975).
55. D. K. Owens, *J. Appl. Polym. Sci.*, *19*: 3315 (1979).
56. A. R. Blythe, D. Briggs, C. R. Kendall, D. G. Rance, and V. J. I. Zichy, *Polymer*, *19*: 1273 (1978).
57. A. Pizzi and N. J. Eaton, *J. Adhes. Sci. Technol.*, *1*: 191 (1987).
58. A. Pizzi, *J. Adhes. Sci. Technol.*, *4*: 573 (1990).
59. A. Pizzi, *J. Adhes. Sci. Technol.*, *4*: 589 (1990).
60. A. Pizzi, *Holzforsch. Holzverwert.*, *43*(3): 63 (1991).
61. D. Levendis, A. Pizzi, and E. Ferg, *Holzforschung*, *46*: 263 (1992).
62. F. M. Fowkes and S. Maruchi, *Org. Coatings Plast. Chem.*, *37*: 605 (1977).

63. F. M. Fowkes and M. A. Mostafa, *Ind. Eng. Chem. Prod. Res. Dev., 17*: 3 (1978).
64. F. M. Fowkes, in *Physicochemical Aspects of Polymer Surfaces*, Vol. 2 (K. L. Mittal, ed.), Plenum Press, New York, 1983.
65. F. M. Fowkes, *Rubber Chem. Technol., 57*: 328 (1984).
66. F. M. Fowkes, D. O. Tischler, J. A. Wolfe, L. A. Lannigan, C. M. Ademu-John, and M. J. Halliwell, *J. Polym. Sci. Polym. Chem. Ed., 22*: 547 (1984).
67. F. M. Fowkes, C. Y. Sun, and S. T. Joslin, in *Corrosion Control Organic Coatings* (H. Leidheiser, ed.), N.A.C.E., Houston, Texas, 1981.
68. J. R. Huntsberger, *J. Adhes., 12*: 3 (1981).
69. F. W. Tischer, U.S. patent 2,177,160 (1939).
70. G. E. Troughton, *Kinetic Evidence for Covalent Bonding Between Wood and Formaldehyde Glues*, Information Report VP-X-26, Forest Products Laboratory, Vancouver, British Columbia, 1967.
71. G. E. Troughton and S.-Z. Chows, *J. Inst. Wood Sci., 21*: 29 (1968).
72. R. M. Rowell and W. D. Ellis, *Wood Sci., 12*(1): 52 (1979).
73. R. M. Rowell, W. C. Feist, and W. D. Ellis, *Wood Sci., 13*(4): 202 (1981).
74. R. M. Rowell and W. D. Ellis, in *Urethane Chemistry and Applications* (K. N. Edwards, ed.), ACS Symposium Series No. 172, Americal Chemical Society, Washington, D.C., 1981.
75. A. J. Morak and K. Ward, Jr., *Tappi, 53*(4): 652 (1970).
76. A. J. Morak and K. Ward, Jr., *Tappi, 53*(6): 1055 (1970).
77. A. J. Morak and K. Ward, Jr., *Tappi, 53*(12): 2278 (1970).
78. W. E. Johns, The chemical bonding of wood, in *Wood Adhesives: Chemistry and Technology*, Vol. 2 (A. Pizzi, ed.), Marcel Dekker, New York, 1989.
79. I. E. Klein, J. Sharon, A. E. Yaniv, M. Doduik, and D. Katz, *Int. J. Adhes. Adhes., 3*: 159 (1983).
80. E. H. Andrews and A. J. Kinloch, *J. Polym. Sci. Symp., 46*: 1 (1974).
81. S. Crisp, H. J. Prosser, and A. D. Wilson, *J. Mater. Sci., 11*: 36 (1976).
82. T. Sugama, L. E. Kukacka, and N. Carciello, *J. Mater. Sci., 19*: 4045 (1984).
83. H. T. Chu, N. K. Eib, A. N. Gent, and P. N. Henriksen, in *Advances in Chemistry Series No. 174* (J. L. Koenig, ed.), American Chemical Society, Washington, D.C., 1979.
84. A. Pizzi, E. P. von Leyser, J. Valenzuela, and J. Clark, *Holzforschung, 47*: 168 (1993).
85. K. Frisch, L. P. Rumao, and A. Pizzi, in *Wood Adhesives: Chemistry and Technology*, Vol. 1 (A. Pizzi, ed.), Marcel Dekker, New York, 1983.
86. A. Pizzi, *J. Polym. Sci. Polym. Chem. Ed., 20*: 707 (1982).
87. A. Pizzi, *Wood Sci. Technol., 17*: 303 (1983).
88. A. Pizzi, *Holzforsch. Holzverwert., 31*: 128 (1979).
89. A. Pizzi, *J. Appl. Polym. Sci., 25*: 2547 (1980).
90. A. Pizzi, *J. Polym. Sci. Polym. Chem. Ed., 19*: 3093 (1981).
91. J. G. Ostmeyer, T. Elder, D. M. Littrell, B. J. Tatarchuk, and J. E. Winandy, *J. Wood Chem. Technol., 8*: 413 (1988).

92. J. G. Ostmeyer, T. J. Elder, D. M. Littrell, B. J. Tatarchuk, and J. E. Winandy, *J. Wood Chem. Technol.*, *8*: 578 (1988).
93. A. Pizzi, *Holzforschung*, *44*: 373 (1990).
94. A. Pizzi, *Holzforschung*, *44*: 419 (1990).
95. A. Pizzi, W. E. Conradie, and A. Jansen, *Wood Sci. Technol.*, *20*: 71 (1986).
96. A. Pizzi and W. E. Conradie, *Mater. Org.*, *21*: 31 (1986).
97. F. A. Cameron and A. Pizzi, *Holz Roh Werkst.*, *43*: 149 (1985).

2

Urea–Formaldehyde Adhesives

I. INTRODUCTION

Urea–formaldehyde (UF) resins are the most important and most used class of aminoresin adhesives. UF resins are polymeric condensation products of the reaction of formaldehyde with urea. The advantages of UF adhesives are their (1) initial water solubility (this renders them eminently suitable for bulk and relatively inexpensive production), (2) hardness, (3) nonflammability, (4) good thermal properties, (5) absence of color in cured polymers, and (6) easy adaptability to a variety of curing conditions [1,2].

Thermosetting amino resins produced from urea are built up by condensation polymerization. Urea is reacted with formaldehyde, which results in the formation of addition products, such as methylol compounds [1]. Further reaction, and the concurrent elimination of water, leads to the formation of low-molecular-weight condensates that are still soluble. Higher-molecular-weight products, which are insoluble and infusible, are obtained by further condensing the low-molecular-weight condensates. The greatest disadvantage of the amino resins is their bond deterioration, caused by water and moisture. This is due to the hydrolysis of their aminomethylenic bond. Therefore, UF adhesives are used only for interior applications.

19

The reaction between urea and formaldehyde is complex and excellent and extensive reviews dealing with it already exist. The combination of these two chemical compounds results in both linear and branched polymers, as well as tridimensional networks, in the cured resin. This is due to a functionality of 4 in urea (due to the presence of four replaceable hydrogen atoms) (in reality only trifunctional), and a functionality of 2 in formaldehyde. The most important factors determining the properties of the reaction products are (1) the relative molar proportion of urea and formaldehyde, (2) the reaction temperature, and (3) the various pH values at which condensation takes place. These factors influence the rate of increase of the molecular weight of the resin. Therefore, the characteristics of the reaction products differ considerably when lower and higher condensation stages are compared, especially solubility, viscosity, water retention, and rate of curing of the adhesive. These all depend to a large extent on molecular weights.

The reaction between urea and formaldehyde is divided into two stages. The first is the alkaline condensation to form mono-, di-, and trimethylolureas [1]. (Tetramethylolurea has never been isolated.) The second stage is the acid condensation of the methylolureas, first to soluble and then to insoluble cross-linked resins. On the alkaline side, the reaction of urea and formaldehyde at room temperature leads to the formation of methylolureas.

On the acid side, the products precipitated from aqueous solutions of urea and formaldehyde, or from methylolureas, are low-molecular-weight methyleneureas [2]:

$$H_2NCONH(CH_2NHCONH)_nH$$

These contain methylol end groups in some cases, through which it is possible to continue the reaction to harden the resin. The monomethylolureas formed copolymerize by acid catalysis and produce polymers and then highly branched and cured networks.

The kinetics of the formation and condensation of mono- and dimethylolureas and of simple UF condensation products has been studied extensively. The formation of monomethylolurea in weak acid or alkaline aqueous solutions is characterized by an initial fast phase followed by a slow bimolecular reaction [3,4]. The reaction is reversible. The rate of reaction varies according to the pH, with a minimum rate of reaction in the pH range 5 to 8 for a molar ratio of 1:1 for urea–formaldehyde and a pH of ±6.5 for a 1:2 molar ratio [5] (Fig. 2.1).

The rapid initial addition reaction of urea and formaldehyde is followed by a slower condensation, which results in the formation of polymers [6]. The rate of condensation of urea with monomethylolurea to form methyl-

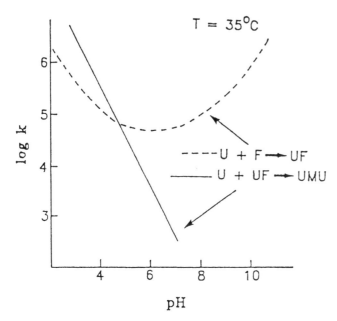

Figure 2.1 Influence of pH on the addition and condensation reactions of urea and HCHO. U, urea; F, HCHO; M, $-CH_2-$.

enebisurea (or UF "dimers") is also pH dependent. It decreases exponentially from a pH of 2 to 3 to a neutral pH value. No condensation occurs at alkaline pH values.

The initial addition of formaldehyde to urea is reversible and is subject to general acid and base catalysis. The forward bimolecular reaction has an activation energy of 13 kcal mol^{-1}. The reverse unimolecular reaction has an activation energy of 19 kcal mol^{-1} [4].

Methylenebisurea and higher oligomers undergo further condensation with formaldehyde [7] and monomethylolurea [8], behaving like urea. The capability of methylenebisurea to hydrolize to urea and methylolurea in weak acid solutions (pH 3 to 5) indicates the reversibility of the aminomethylene link and its lability in weak acid moisture. It explains the slow release of formaldehyde over a long period in particleboard and other wood products manufactured with UF resins.

It is very important in the commercial production of UF resins to be able to control the size of the molecules by the condensation reaction, since their properties change continuously as they grow larger. The most perceptible change is the increase in viscosity. Low-viscosity syrups are formed first. These change into high-viscosity syrups that are clear to turbid.

The most common method for the preparation of commercial UF resin adhesives is the addition of a second amount of urea during the preparation reaction. This consists of reacting urea and formaldehyde in more than equivalent proportions. Generally, an initial urea/formaldehyde molar ratio of 1:2.0 to 2.2 is used. Methylolation can in this case be carried out in much less time by using temperatures of up to 90 to 95°C. The mixture is then maintained under reflux.

When the exotherm subsides (usually after 10 to 30 min), the methylol compounds have formed, and the reaction is completed under reflux by adding a trace of an acid to decrease the pH to the UF polymer-building stage (pH 5.0 to 5.3). As soon as the right viscosity is reached, the pH is increased to stop polymer building and the resin solution is cooled to about 25 to 30°C. More urea (called second urea) is added to consume the excess of formaldehyde until the molar ratio of urea to formaldehyde is in the range 1:1.1 to 1:1.7. After this addition of urea, the resin is left to react at 25 to 30°C for as long as 24 h. The excess water is eliminated by vacuum distillation until a resin solids concentration of 64 to 65% is reached, and the pH is adjusted to achieve suitable shelf life or storage life.

II. MOLECULAR DISTRIBUTION AND EFFECT ON RESIN AND BOARD PROPERTIES

UF resins can and have been synthesized at higher molecular mass. It has been found that weight-average molecular mass increases with longer condensation at the turbidity temperature [9] from several thousands up to more than 100,000 g mol^{-1}. These high-condensation UF resins were also found to contain portions with molar masses up to 500,000 g mol^{-1} which are not only aggregates caused by physical intermolecular forces [9]. The higher-molecular-mass portions are present at considerably higher concentration in the aqueous disperse phase than in the aqueous solution phase [10].

At a higher formaldehyde/urea ratio, the higher is the content of free formaldehyde. Comparing values given by various authors shows an exponential increase of free formaldehyde with increasing formaldehyde/urea molar ratio. Such differences in volume as observed might also be caused by differences in condensation type, preparation conditions, and so on, as has been critically reviewed by Kasbauer et al. [11]. A higher content of free formaldehyde enables, on the one hand, promotion of acid by reacting with the ammonium chloride hardener: in this way the hardening is also correlated significantly to free formaldehyde left present and hence to formaldehyde emission [12].

Reaction between urea and formaldehyde leads to a steady state, which in the simpler case of the methylolation of urea is representable by an equilibrium constant dependent on formaldehyde supply, concentration of reactants, and temperature [12]:

$$K = \frac{k_1}{k_2} = \frac{[UF]}{[U][F]}$$

In the case of condensed resins, a steady state exists between free formalde-hyde, N-methylolformaldehydes (at the end of chain as a mono- or dimethylol group or as a methylol group at a branching site), and the remaining amido groups ($-NH_2$, $-NH-$, unreacted urea). This steady state can be described by quantitative determination of the contents of different structural groups in the resin by combining chemical analysis and NMR [12–15].

The influence of the formaldehyde content on gelation and hardening of the resin is shown by the dependence of the required gel time on the content of free formaldehyde [12] (Fig. 2.2). Above 0.2% free formaldehyde there is slight dependence of gel time on free formaldehyde for salts such as ammonium chloride and monoammonium phosphate, whereas when using an acid as the hardener, the gel time is nearly constant [12]. On the other hand, gel time lengthens significantly for resins with less than 0.2% free formalde-hyde. In these resins monoammonium phosphate at 100°C does not gel the resin even after 5 min. This 0.2% free formaldehyde limit is also recognizable by a plot of the inverse of the gel time at 100°C as a function of the free formaldehyde content. This parameter is a direct reflection of the reaction rate under the condition used for gelation [12].

For example, using ammonium chloride as hardener, the curing reaction rate increases linearly above 0.2%, whereas under this limit there is a noticeable decrease. A similar trend is noticeable when correlating gel time with the amount of ammonium chloride hardener used. The shape of such a curve is practically independent of the type of resin used. Additions of ammonium chloride below 1.5% give noticeable lengthening of gel times. Additions higher than 1.5% do not yield any significant shortening of the gel time [12].

An increase in molar mass and a decrease in the formaldehyde/urea molar ratio tend to result in increased swelling and water absorption [12]. This molar mass influence is more pronounced for resins of lower molar ratio [12].

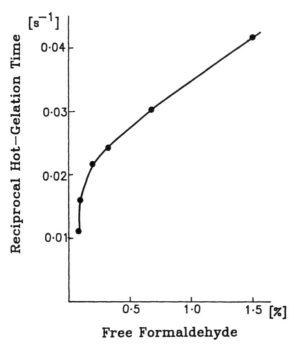

Figure 2.2   Reciprocal of gelation time as a function of free formaldehyde for UF resins.

Bending strength was found to be relatively independent of molar mass distribution on molar ratio [12]. By contrast, tensile strength instead appears to decrease to some extent when using resins of lower molar ratio [12]. Subsequent emission of formaldehyde increases noticeably with increasing molar ratio but appears to be independent of molar mass distribution.

III. QUANTITATIVE CHARACTERIZATION OF UF ADHESIVE PERFORMANCE BY ^{13}C NMR

Characterization of the structure of UF resins by ^{13}C NMR of their solutions is a well-established and important technique [15–19]. It is the best method at the molecular level of characterizing and documenting the variations in resin chemistry for UF adhesives. Figure 2.3 shows the ^{13}C NMR spectrum of a UF resin on which the various bonds are assigned to well-defined chemical groups in the resin. It is possible just by the appearance of the spectrum to

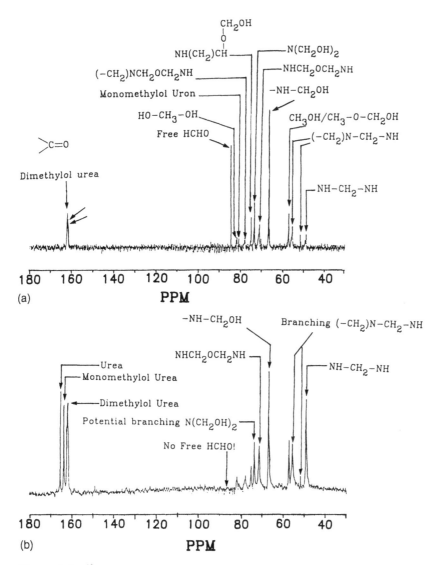

Figure 2.3 ^{13}C NMR spectra of (a) a high-strength, low-formaldehyde-emission UF resin (U/F molar ratio 1:1.8), and (b) a low-strength, low-formaldehyde-emission UF resin (U/F molar ratio 1:1).

deduce quantitatively some of the characteristics of the resin. For example, the high-intensity bond at 84 ppm, indicative of free formaldehyde, indicates that this is a high-emission resin (Fig. 2.3a).

^{13}C NMR monitors characteristics of the resin at the molecular level. Such characteristics are supposed to determine the macro properties of the resin and the properties of particleboard bonded with it. It should then be possible to relate the ^{13}C NMR spectrum of a liquid UF resin quantitatively to the performance of the board bonded by the same, hardened resin. Evolution of the cured strength of a UF adhesive in a board and the determination of its capacity for formaldehyde emission, once it has become the hardened binder of the board, is a long procedure that necessitates repetitive testing, as it is subject to fairly large experimental errors. These are induced by the need to pass through the board manufacturing stage. Recently, however, a method based on the use of ^{13}C NMR relative peak intensities ratios for different characteristic chemical groups, known or supposed to contribute to UF resins strength and formaldehyde emission, has been developed [20,21]. The absolute intensities of ^{13}C NMR peaks can never be taken as a measure of the abundance of a particular chemical group. In a resin such as UF in which all chemical groups are closely interrelated, ratios of the integrated intensities of peaks characteristic of chemical groups which are known or suspected to contribute to the cured strength and formaldehyde emission appear to indicate excellent correlation with the experimental reality.

Such an approach was pioneered by Tomita and Hatono [15], who devised a simple equation relating ^{13}C NMR band intensities ratios to resin cross-linking. The equation did not go as far as to relate ^{13}C NMR and intensities ratios to the internal bond strength of a particleboard bonded with the UF resin. The equation described the phenomenon well but was found later to be able to describe the cured strength of the resins only up to a level of approximately 75% significance [20,21].

A much later study [20,21] obtained a more complete equation, in which the Tomita and Hatono equation still constituted one of the main terms. This general equation relating the extent of cured resin cross-linking and thus strength to certain ^{13}C NMR peak ratios is

$$\text{resin cross-linking} = \frac{C+2E}{A+C+E} + \frac{urea}{C1+C2} + \frac{Me}{Mo}$$

[20,21] and therefore for the internal bond (IB) strength of a particleboard made under well-defined conditions,

$$\text{IB strength} = a\frac{C+2E}{A+C+E} + b\frac{\text{urea}}{C1+C2} + \frac{Me}{Mo}$$

where a, b, and c are coefficients characteristic for each unique homologous series of UF resins, at an average density of 0.680 g cm^{-3}; $A = -NH_2 \text{ } CH_2NH-$ (peak at 48.8 ppm); $C = -N(CH_2)CH_2NH-$ (peak at 55.5 ppm); $E = -N(CH_2-)CH_2N(CH_2^-) -$ (peak at 61.6 ppm); $C1 = NH_2CONHCH_2OH$ (peak at 163.0 ppm); $C2 = $ larger molecular species [i.e. $= NCON = $ groups ($HOCH_2NHCONHCH_2OH$) (peak at 162.0 ppm)]; urea = free urea (peak at 165 ppm); $Me = A + C + E$, total methylene species (peak from 45 to 60 ppm); and $Mo = $ total methylol species (peak from 65 to 72 ppm).

Two series of resins, each series using a different system of manufacture, and each resin within a series of varied urea–formaldehyde molar ratio, were tested and the relevant a, b, and c coefficients derived. The two series for which the coefficients were determined were (1) a series of standard UF resins produced by second urea addition, and (2) a series of resins produced by a higher and progressive number of subsequent urea additions. The a, b, and c coefficients for the first series were 0.4275, –0.3623, and 2.389, and 1.077, –0.1130, and –0.1424, respectively, for the second series of resins. These for the urea/formaldehyde molar ratio range from 1:1.8 to 1.0.9 for boards of 0.680 g cm^{-3} [20,21].

A similar approach has been taken for relating ^{13}C NMR peak ratios of different chemical groups in the liquid UF resin to the formaldehyde emission of particleboard bonded with it [20,21]. An equation has been obtained that is unique for a specific homologous series of UF resins. This is also unique to the particular formaldehyde emission determination method (desiccator) [20,22], although this is easily correlated with other methods [20,23,24], and in the form given below the results are given in the WKI method as milligrams per 100 grams of board. This general equation linking the ^{13}C NMR intensity ratios of free F/Me, Tomita and Hatono cross-linking ratio, urea/(C1 + C2), (E1 + E2)/(E4 + E5), and E3/Me to the formaldehyde emission of a board can be expressed as

$$\text{F emission} = a\frac{\text{free F}}{Me} + b\frac{C+2E}{A+C+E} + c\frac{\text{urea}}{C1+C2} + d\frac{E1+E2}{E4+E5} + e\frac{E3}{Me}$$

where a, b, c, d, and e are coefficients characteristic for each unique homologous UF resin series and the formaldehyde emission is expressed in mg HCHO per 100 g of board; free F = free formaldehyde (peak at 84.7 ppm); Me = sum of all methylene groups; A = $-NHCH_2NH-$ (peak at 48.8

ppm); C = $-N(CH_2-)CH_2NH-$ (peak at 55.5 ppm); E = $-N(CH_2-)CH_2N$ (CH_2-) (peak at 61.6 ppm); C1 = $NH_2CONHCH_2OH$ (peak at 162.0 ppm); C2 = larger molecular species [i.e. = $NCON$ = groups (peak at 162.0 ppm)]; urea = free urea (peak at 165 ppm); E1 = $-NHCH_2OCH_3$ (peak at 73.4 ppm); E2 = $-N(CH_2CH_2OCH_3$ (peak at 88.7 ppm); E4 = $-NHCH_2OCH_2NH-$ (peak at 70.3 ppm); E5 = $-N(CH_2-)CH_2OCH_2NH-$ (peak at 75.0 ppm); E3 = $-NHCH_2OCH_2OH$ (peak at 88.7 ppm).

Most of the ratios chosen appear to correlate with the formaldehyde emission. As clearly expected, the free F/Me ratio correlates excellently with the formaldehyde emission from particleboard. The effect that other group ratios, such as C1/C2, have on the final emission of formaldehyde from the board is less clear. An equation relating only the free F/Me ratio to the formaldehyde emission can also be obtained for the resins [20,21]. This ratio would only be applicable for shorter periods of time since it has been shown that the formaldehyde emission from boards made with a resin that initially contained high amounts of free formaldehyde in the liquid resin shows a considerable decrease when tested again after 3 months.

The correlations of formaldehyde to the other NMR peak ratios were also reasonably good. Thus they could also be considered to play a role in the formaldehyde emission of particleboards and be included in a final equation that expresses the formaldehyde emission of a UF-bonded board to the ^{13}C NMR peak ratios of the groups present in the liquid resin.

The formaldehyde/urea molar ratio does, of course, correlate very well with the formaldehyde emission of the particleboard [20,21]. This is not, however, included in the final equation because that implies knowledge of this ratio. It is therefore useful when the ratio is known, but it is of no use when a resin of unknown ratio needs to be evaluated.

The urea/(C1 + C2) ratio should logically belong to the equation, as the presence of free urea in the system would affect the final formaldehyde emission by "capturing" free formaldehyde during the curing stages of the resin, with the urea being more efficient but more random in this effect than the C1 and C2 species. The correlation of the C1/C2 ratio seems to be very good, but its contribution to the final formaldehyde emission of a board is not clear. Since the carbonyl ratios have already been considered indirectly in the urea/(C1 + C2) ratio, one of the two ratios is likely to be redundant. The latter is likely to be the C1/C2, as the urea/C1 + C2 takes into consideration not only C1 + C2 but also the free urea.

The cross-linking ratio [(C + 2E)/(A + C + E)] from Tomita and Hatono should also have some influence on the formaldehyde emission of a particleboard. This is especially true for boards in service for long periods of time and exposed to higher-humidity conditions. Since, with time, most of the free formaldehyde, which influences the emission to a significant degree, has almost all disappeared, the $-CH_2-$ groups present then play a role. Hence an increase in cross-linking of a resin means that there is a higher concentration of $-CH_2-$ groups present, which with time might slowly hydrolyze and emit formaldehyde.

The next ratios to consider are those due to the ethers. The correlation of all three ratios is good, but only one of the ratios can be considered. This is because looking at the chemistry of each of the ethers present, one must consider which would contribute to the formaldehyde emission. The E1/E2 ratio (where E1 = $-NHCH_2OCH_3$) would not influence the emission since the species would simply decompose and emit methanol rather than formaldehyde. The (E1 + E2)/(E4 + E5) species ratio was the one preferred to describe the contribution of the ethers to the formaldehyde emission, even though the correlation of the second ratio appeared at first to be slightly better. There is a well-defined experimental reason for this: E4 = $-NHCH_2OCH_2NH-$ and E5 = $-N(CH_2-)CH_2OCH_2NH-$. These species can decompose and emit formaldehyde as follows:

$$-NHCH_2OCH_2NH- \rightarrow -NHCH_2NH- + \text{formaldehyde}$$

Hence a decrease in the abundance of the (E4 + E5) species is likely to decrease the emission of formaldehyde in the cured resin. Only one ether ratio is then necessary to describe the effect, as to use all three ratios would mean to consider the same or similar effects to contribute three times and give a false impression of the importance of the ethers to the emission. The coefficients a, b, c, and d for the two series of resins tested were 19.18, -11.66, 7.166, and 4.126, respectively, for the simpler resin series and 21.71, -6.086, 2.389, and 4.346 for the multiple-step urea addition series [20,21].

A further indication of the effects indicated with regard to the strength of the cured UF resin was obtained by a follow-up study by x-ray diffraction of the UF resins prepared after they were cured [20,21]. The crystallinity of the hardened resins changes from an almost amorphous material (UF = 1:1.8) to a semicrystalline material (UF = 1:0.7). The proportion of crystallinity of the respective resins cured alone is summarized as follows [20,21]:

Increase in urea molar proportion					
1:1.8	1:1.5	1:1.3	1:1.1	1:0.9	1:0.7
Crystallinity 17%	21%	24%	43%	47%	74%

The two methods of synthesis had no apparent effect on the final cured resin crystallinity [20,21]. Previous studies [25] showed that the crystallinity of UF resins disappears when cured on wood, this fact being supported by previous theoretical considerations [26]. This also occurred for the resins under examination. For this reason the crystallinity of the hardened UF resins were determined by x-ray diffraction only to present additional evidence of why the strength of the final resin is related to the urea/formaldehyde molar ratio.

This suggests that the crystallite regions of the resin are not separate from the bulk amorphous region but rather, consist of small regions where crystalline order is formed by sections of the molecules [25]. These are regions where $-CH_2-$ bridges linking tridimensionally single chains of resin do not occur and thus secondary forces adjust the molecules in a crystalline state [25]. There should then be some correlation between the extent of crystallinity of the resin when cured alone and the strength of the same UF resin when used as an adhesive. The higher the percent crystallinity, the lower the amount of tridimensional cross-linking, and hence the weaker the strength of the hardened resin should be.

Thus an attempt was made to correlate IB strength and percent crystallinity. The results in Table 2.1 indicate that percent crystallinity is correlated primarily with the urea/(C1 + C2) ^{13}C NMR peak intensities ratio, and less well with the (C + 2E)/(A + C + E) ^{13}C NMR ratio. The influence of the latter term is expected because as the tridimensional cross-linking of the resin decreases, the percent crystallinity should increase (the proportion of linear species increases), allowing for a higher degree of ordered packing to occur. Similarly, the influence of the urea/(C1 + C2) term is expected because as

Table 2.1 ^{13}C NMR Species Ratio in the Liquid Resin as Related to the Percent Crystallinity of the Hardened UF Resin

NMR species ratio	Correlation coefficient
Urea/(C1 + C2)	0.941
C1/C2	0.621
E1/E2	0.569
Cross-linking (C + 2E)/(A + C + E)	0.800
Me/Mo	0.439
(E1 + E4)/(E2 + E5)	0.631
(E1 + E2)/(E4 + E5)	0.542
Free F/Me	0.759

the monomeric urea and mono- and dimethylol ureas increase, the proportion of linear species increases and the percent crystallinity should also increase. As a consequence, the following equation correlating crystallinity to ^{13}C NMR species ratios was proposed [20,21]:

$$\% \text{ crystallinity } + 25.75\frac{\text{urea}}{\text{C1}+\text{C2}} + 40.64\frac{\text{C}+2\text{E}}{\text{A}+\text{B}+\text{C}}$$

The equation above presents two of the terms relating IB strength with ^{13}C NMR peak ratios, clearly indicating that a correlation exists between resin strength and percent crystallinity. Interesting, and expected, is the fact that in the equation above the urea/(C1 + C2) term's coefficient is opposite in sign to that in the strength equation, indicating that the contributions of these species to strength and percent crystallinity are inversely proportional.

IV. MOLECULAR MECHANICS OF UF RESIN ADHESION TO CELLULOSE SUBSTRATES

A. Fundamental Aspects

Studies of the three-dimensional space conformation of UF resins have been carried out successfully [27–29]. However, it has also been determined that when taken in isolation, the configuration of polymeric resin oligomers does not correspond to their minimum energy configuration when interacting with the surface of a substrate [1,30]. This indicates that while studies of the internal UF resin configurations taken in isolation are of considerable importance to the chemistry of UF resins, they are less important in the chemistry of the resin/cellulose substrate interface.

The finding that the configuration of the polymeric resin molecule has a marked influence on its energy of absorption on a particular substrate, and hence on adhesion, is of some applied as well as theoretical interest. The molecular mechanics approach can indicate how the preparation of the adhesive could be changed to improve its adhesion to a well-defined substrate.

Relevant molecular mechanics studies of interfacial adhesion between UF resins and amorphous and crystalline cellulose exist [26,31]. In these studies the energies of adhesion of a series of UF oligomers with the substrate were obtained, and the results were found to be relevant to an understanding of the level of adhesion to lignocellulosic substrates in relation to the UF formulation used [26,31,32].

The results indicate that there are significant differences in the values of

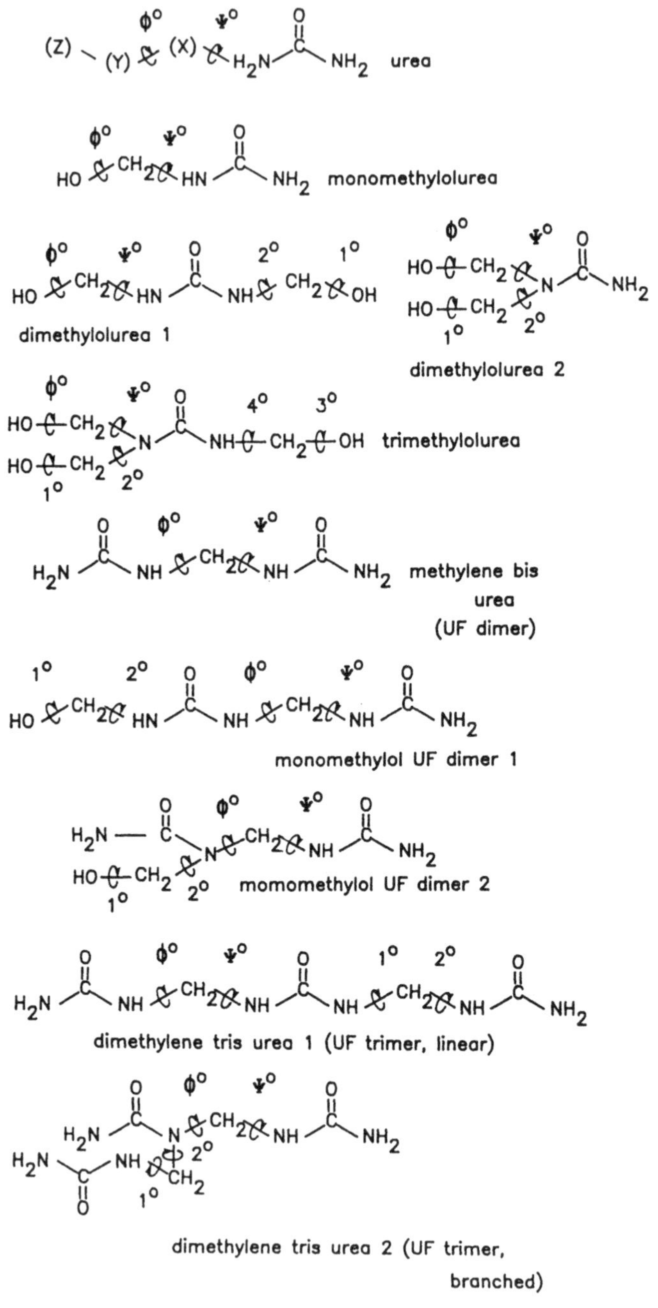

Figure 2.4 UF oligomers studied by molecular mechanics techniques.

the minimum total energy in the interaction of various UF condensates and crystalline and amorphous cellulose. The UF species examined show attraction to, and hence adhesion to, the surface of the two types of cellulose. The adhesion of UF resins, however, appears to follow different patterns according to the predominance of the chemical species present in the resin used. The series urea–monomethylolurea–dimethylolureas shows that, in general, adhesion increases with increasing amount of methylol groups in the resin, except when steric hindrance is also increased. This is confirmed by the difference in specific adhesion to cellulose between the UF dimer and its linear monomethylol derivative (Fig. 2.5 and 2.6). The effect of resin branching, and hence of steric hindrance, is noticeable at several stages. The branched dimethylolurea, branched monomethylol UF dimer, and branched UF trimer interactions with the cellulose surfaces show a marked drop in adhesion from their linear counterparts. The same is noticeable for trimethylolurea, where the opposite effects of further methylolation and increased steric hindrance lead to adhesion lower than that for linear dimethylolurea but higher than that for branched dimethylolurea.

As the reactivity toward formaldehyde of an amide group already carrying a methylol group is lower than that of an amide group with no methylol group on it by a factor of approximately 1:3 [1,33], the two possible dimethylolureas should most commonly be in the proportion of 3:1 in favor of the linear isomer; the branched dimethylolurea is thus, in general, about 25% of the total dimethylolureas formed. The average specific adhesion to crystalline cellulose I of the mixture of the two methylolureas is then still higher than that of monomethylolurea (approximately -15 to -14 kcal mol^{-1}). Similar reasoning can be applied to the other methylolated species.

As regards improved adhesion, it does appear that, in general, methylolation has a beneficial effect. There is considerable experimental evidence to support these theoretical findings [34–37]. First, UF resins in which the ratio of formaldehyde to urea is higher tend to give better adhesion [13,16]; second, urea addition to end a UF resin preparation, used to mop up free formaldehyde still present, is well known to improve the adhesion of the resin [15]. It is clear that these effects are obtained in the first case by the increased amount of methylol groups on the UF oligomers, and in the second case by the formation of methylolureas by terminal urea addition to the cooled UF resin. Inoue and Kawai [35], for instance, found that the adhesion strength appears to be at its maximum at a low degree of condensation of the UF resin: the theoretical findings by molecular mechanics confirm this deduction. While methylol groups in UF resins have been found to form covalent ether links with cellulose [38,39], the theoretical findings also indicate clearly that the

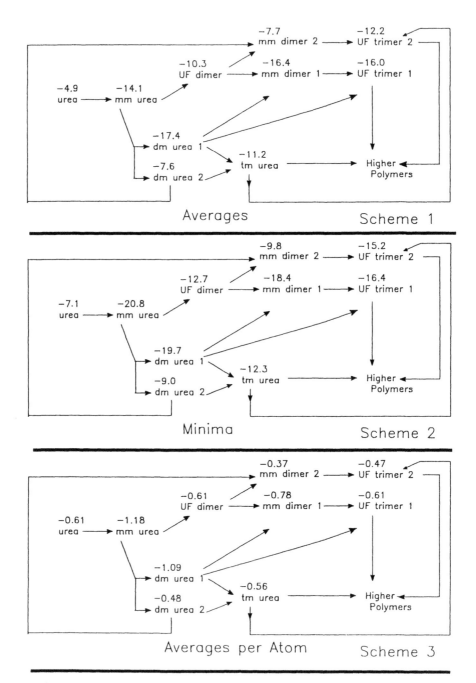

Figure 2.5 Averages, minima, and averages per atom of the interaction energy (energy of adhesion) of UF oligomers with crystalline cellulose I (in kcal mol^{-1}).

Scheme 1

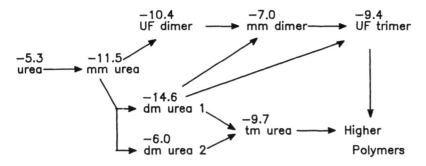

Minima

Scheme 2

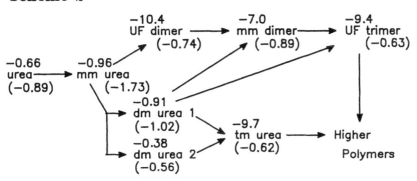

Minima : Averages Per Atom

* ( ) value for crystalline cellulose I

Figure 2.6 Minima and averages per atom of the interaction energy (energy of adhesion) of UF oligomers to amorphous cellulose (and crystalline cellulose I in parentheses), in kcal mol^{-1}).

sum of secondary forces gives equally good or better bonding. Secondary forces can then give excellent adhesion without the occurrence of covalent bonds, as asserted by other authors [40].

The variations in specific adhesion of the nonmethylolated series urea–UF dimer–UF trimer are also of interest. Their minimum average energy and minimum energy of adhesion per mole of each chemical species increase when progressing from urea to UF dimer to UF trimer (Fig. 2.5, schemes 1 and 2). The increase is again more marked with linear rather than branched oligomers. The effect of branching in the molecule is also of particular interest: to set and cure a resin means to introduce branching to form a three-dimensional cross-linked network. If one extrapolates from the low-molecular-weight species examined, the lower energies of adhesion of branched UF species indicate that on curing there should be a decrease not of the total attractive forces, but at least of the efficiency of adhesion per atom of the molecule. Thus while the average adhesion of a branched UF trimer is 12.2 kcal mol^{-1}, higher than that of a linear dimer at 10.3 kcal mol^{-1}, the efficiency of adhesion for each atom of the molecule (i.e., the total adhesion energy divided by the total number of atoms in the molecule) decreases from the linear UF dimer at 0.61 kcal mol^{-1} to the branched UF trimer at 0.47 kcal mol^{-1}. The concept of efficiency of adhesion per atom is of interest. The efficiency of adhesion is exactly the same (0.61 kcal mol^{-1}) for urea, UF dimer, and UF trimer (Fig. 2.5, scheme 3), indicating that for linear-growing UF polymers its value appears to remain constant. The conclusion that can be drawn is that while adhesion itself appears to be a direct function of molecular size, in linear UF polymers adhesion efficiency to cellulose appears to be constant.

On comparing the interaction of UF oligomers with cellulose with the water adsorption of crystalline cellulose I, it is evident that while UF oligomers and methylolureas present more attractive interactions than the average per water molecule (5.5 kcal mol^{-1} [20]), these attractive interactions are lower for UF resins than for those observed for phenol–formaldehyde (PF) dimers on the same sites [30]. Urea itself has lower adhesion (4.3 kcal mol^{-1}) than the average adhesion per water molecule, indicating that it is more easily displaced by water on most of its adsorption sites on cellulose. While PF dimers have an average adhesion energy generally higher than that of water molecules, even at the stronger cellulose sorption sites [41], the UF oligomers (with the exception of some methylolureas) have an average adhesion value between the average and maximum energy of adsorption of water on cellulose.

It must be noted here that in this study [26,30,31] all the calculations were carried out using the dielectric constant of water to lower the electrostatic energy component. This is important, as it means that adhesion of the resin (UF and PF) to the cellulose surface is obtained by secondary forces and electron transfer effects even when molecules of water (only a "monolayer," however [41]) are present on the cellulose surfaces. Thus, by comparing the energy of adhesion of PF and UF resins, it appears that PF resins [30] are much more tolerant of water and moisture content than are UF resins in terms of adhesion through secondary forces with the crystalline cellulose I surface. This is in agreement with experimental evidence [1,33]. Use of the dielectric constant of water in the calculations was introduced to take into account the fact that water, which is the resins' carrier, as well as being present in the cellulose and wood, will interfere with the process of adhesion, modifying the secondary forces between the adhesive and the adherend.

A further conclusion implied is that lack of water and weather resistance of lignocellulosic materials bonded with UF resins does not appear to be due, to any large extent, to failure in their adhesion to cellulose (on average, this is better than that of water) although this appears to be a contributory factor. This implies that it is the decomposition under moisture attack of the aminoplastic bond within the resin itself that causes its lack of water resistance. This is also supported by circumstantial experimental evidence [23].

An interesting difference between PF [1] and UF resins is that in the latter the importance to adhesion of hydrogen bonding is much lower than for phenolics. This is due to the predominance of strong intramolecular hydrogen bonds in the UF oligomers even when the molecules are removed from their internal configuration of minimum total energy.

Another interesting conclusion implied by molecular mechanics studies [26,31] is that when UF resins are interacting with cellulose, some of the conformations that would be forbidden when the UF resin is cured alone become possible and are allowed. This will militate against at least the first layer of UF oligomers in contact with cellulose being in a microcrystalline state. Although it appears from the results of other authors [28] that some UFs in isolation are in a microcrystalline state, this is not the case for the first few layers of resin on the cellulose surface, where crystallinity dislocations due to other now-favored UF conformations are present. In thick glue lines, it may well be that after the first few layers the resin reverts to the microcrystalline structures advocated [28]. In close-contact wood bonding, however, where glue lines are very thin, UF resins are more likely to be in amorphous or possibly paracrystalline forms rather than being microcrystalline.

This deduction has been confirmed by x-ray diffraction studies of UF resins cured in the presence of wood [25]. These showed experimentally that when cured in the presence of wood, even when the proportion by mass of solid resin to solid wood is as high as 50:50, a UF resin does not present any crystallinity. This means that either the solid resin cured on wood is amorphous, or its crystallinity is so low that it cannot be detected. This experimental study [20,21] also confirmed that the secondary forces binding together linear chains, not cross-linked, of UF oligomers with cellulose are stronger than the intermolecular forces between the UF oligomers. This is probably the reason why when the UF is a good resin, it is always likely to be in an amorphous state when cured on wood. It also implies that if a UF-bonded joint in higher-crystallinity, low-molar-ratio resins has to fail, it is likely to fail between the resin molecules or in the wood rather than at the interface between UF resin and cellulose.

The secondary interactions of urea and UF oligomers with amorphous cellulose are, on average, weaker than those with crystalline cellulose I. Results indicate that in several positions the sum of steric hindrance and repulsive secondary forces causes amorphous cellulose to have a higher affinity for water (on average 6.5 kcal mol^{-1}) [30,41] than for UF oligomers.

Furthermore, if one considers that the distribution of water sorption energies is much narrower in amorphous [30,42] than in crystalline cellulose I [26,41], good UF adhesion to amorphous cellulose can still be achieved, but only on selected sites. Adhesion in this case is more site specific than in crystalline cellulose I. Although this is at first surprising, it is not really so if one considers that amorphous cellulose has a considerably more open structure and thus sorbs more water more readily than does crystalline cellulose I [42,43]; at equivalent relative humidities, the equilibrium moisture content is much higher for amorphous than for crystalline cellulose I [44]. Both these factors militate against good UF adhesion, and with the exception of the more selective substrate sites, removal of the UF resin by water is more likely in a more open structure, such as amorphous cellulose.

While in most industrial processes the percentage of UF resins to cellulose is small and the highly attractive sites selected are likely to give more than adequate adhesion, an increase in the amount of UF resin should not lead readily to a comparable increase in adhesion when dealing with amorphous cellulose (and possibly hemicelluloses). The reverse is likely to apply to crystalline cellulose I. It appears that it will be the balance between crystalline and amorphous carbohydrates in the substrate that is likely to determine which effect, increased adhesion or not, is noticeable at the macrolevel.

The first important concept that emerges from the discussion above is that wetting of a substrate by an adhesive can be independent of its solvent and water contents. In the case under examination, water is more a liability than of an aide to UF adhesion by secondary forces. Second, it appears likely that the major part of the susceptibility to water attack in UF resin adhesion resides in the amorphous region of the substrate, at least in the amorphous carbohydrate region (for adhesion only). At this stage, one should start speculating if the covalent bonding of UFs with cellulose reported by other authors [39], particularly in the textile field, is likely to be formed in regions where secondary attractive forces are poorer and where water is more easily adsorbed. As in textiles, wood crystalline cellulose II is considerably more prone than crystalline cellulose I to water sorption [45]. The higher water content in crystalline cellulose II and in amorphous cellulose might well favor ionic reactions to form covalent bonding; UF adhesion then might well be the sum of covalent bonding and secondary forces in these cases. This, of course, would not be valid for wood, where crystalline cellulose I will militate against ionic reactions.

B. Applied Consequences

From the values of adhesion calculated theoretically (in kcal mol^{-1}) it is also possible to determine the specific adhesion of a UF resin to natural cellulose (i.e., the sum of amorphous and crystalline cellulose) according to the relative abundance of the various UF oligomers in the resin. Diagrams such as those presented in Fig. 2.6 can be used to determine by theoretical means the relative advantages of the various stages of condensation of a UF resin as regards its adhesion capability. The values obtained are not absolute but exclusively comparative; they can be used only to compare one resin oligomer mix with another. In the example in Fig. 2.7 a resin should have an approximate total relative adhesion given by

$$
\begin{aligned}
\text{adhesion} =\ & \frac{\%\ \text{urea}}{100} \times (0.3E_{\text{tot.amporph}} + 0.7E_{\text{tot.cryst}}) \\
& + \frac{\%\ \text{mm urea}}{100} \times (0.3E_{\text{tot.amorph}} + 0.7E_{\text{tot.cryst}}) \\
& + \frac{\%\ \text{dm urea}}{100} \times \{\ [0.3(0.75E_{\text{tot.amorph(dmu1)}} + 0.25E_{\text{tot.amorph(dmu2)}})] \\
& + [0.7(0.75E_{\text{tot.cryst(dmu1)}} + 0.25E_{\text{tot.cryst(dmu2)}})]\} \\
& + \frac{\%\ \text{tm urea}}{100} \times (0.3E_{\text{tot.amorph}} + 0.7E_{\text{tot.cryst}}) \\
& + \frac{\%\ \text{polymers}}{100} \\
& \quad \times [0.3(0.75E_{\text{tot.amorph(UF polymers)}} + 0.7E_{\text{tot.cryst(UF polymers)}}]
\end{aligned}
$$

$$(2.1)$$

Although it is possible to calculate all the terms of such an equation, a problem arises in the polymer term, whose composition is largely unknown. An easier approach would be to solve the equation as a function of the efficiency of adhesion, as, at least for crystalline cellulose I, this is a constant if average energies are considered, and its variations tend to zero with increasing degree of polymerization if minimum energies are considered. Thus for the resin in Fig. 2.7, considering the relative percentages of the five chemical species at condensation stages 1 and 2, we have

condensation stage $1 = 0 + 0.03(0.3 \times 0.96 + 0.7 \times 1.73)$
$+ 0.24(0.3(0.75 \times 0.91 + 0.25 \times 0.38)$
$+ 0.7(0.75 \times 1.02 + 0.25 \times 0.56)]$
$+ 0.11(0.3 \times 0.49 + 0.7 \times 0.62)$
$+ 0.65(0.3 \times 0.36 + 0.7 \times 0.63) = 0.65$

condensation stage $2 = 0.57$

From the above (which is only an example) it appears that as regards adhesion, and adhesion alone, condensation stage 1 gives about 15% better adhesion to natural cellulose than does condensation stage 2. If only a value for adhesion to crystalline cellulose I is needed, all the amorphous terms can be discarded, and vice versa. For the example, the values used in the equation were those of the trimer in Fig. 2.6, scheme 2.

The system described above can give a resin formulator an approximate idea of the conditions necessary to improve the adhesion of a UF resin to lignocellulosic materials. Although it constitutes simply a relative and rapid test, it may be of some use in UF resin applications. If extrapolation to the total amorphous and crystalline carbohydrates of wood must be considered, the coefficients 0.3 and 0.7, indicating the relative proportions of amorphous and crystalline cellulose I, can be changed to 0.5 and 0.5 for softwoods and to 0.53 and 0.47 for hardwoods. It must be stressed that such a system gives an idea of only relative specific adhesion, as factors such as the viscosity of the resin, speed of penetration, and so on, cannot be taken into account by a system as simplified as the one presented.

By fitting the data in Figs. 2.5 and 2.6 and Table 2.2 to equation (2.2), the effect of a UF resin formulation on its adhesion to, for instance, a softwood of 70% cellulose crystallinity and 25% lignin content can be predicted:

$$\text{adhesion} = \frac{\% \text{ urea}}{100} \times (0.3E_{\text{amorph}} + 0.7E_{\text{cryst}}) + \frac{\% \text{ mm urea}}{100}$$

$$\times (0.3E_{\text{amorph}} + 0.7E_{\text{cryst}}) + \frac{\% \text{ dm urea}}{100}$$

$$\times [0.3(0.75E_{\text{amorph(dmu1)}} + 0.25E_{\text{amorph(dmu2)}})$$

$$+ 0.7(0.75E_{\text{cryst(dmu1)}} + 0.25E_{\text{cryst(dmu2)}})] + \frac{\% \text{ tm urea}}{100}$$

$$\times (0.3E_{\text{amorph}} + 0.7E_{\text{cryst}}) + \frac{\% \text{ UF polymers}}{100}$$

$$\times (0.3E_{\text{amorph(polymers)}} + 0.7E_{\text{cryst(polymers)}}) \qquad (2.2)$$

Table 2.2 Average Theoretical Values of the Energy of Adhesion per Atom of UF Oligomers for Urea and UF Condensates

Chemical species	Average energy of adhesion (kcal mol^{-1} per atom of UF species) to:	
	Amorphous holocellulose	Crystalline cellulose
Urea	0.66	0.61
Mononethylolurea	0.96	1.18
N,N'-Dimethylolurea, linear	0.91	1.09 ⎫
		⎬ 0.94
N-Dimethylolurea, branched	0.38	0.48 ⎭
Trimethylolurea	0.49	0.56
Methylenebisurea (UF dimer)	0.61	0.61
Dimethylenetrisurea (UF trimer)	0.36	0.61

Source: Refs. 26 and 31.

As adhesion is a surface physicochemical phenomenon similar to water absorption, the contribution of the lignin component to wood adhesion is likely to be as small as it is in water absorption [44]. Skaar [44], for instance, presents proof that the contribution of lignin to the wood–water isotherm is not greater than 5 to 6%. This is equivalent to saying that the contribution of lignin to wood wetting, a key step in any resin-specific adhesion to lignocellulosic materials, depends on lignin presence only to a level of 5 to 6%. For this reason the lignin contribution to adhesion can be disregarded, because the contribution of other factors is much greater.

A "standard" UF resin preparation in which the total amounts of both urea and formaldehyde are added initially to the reaction vessel shows variation in the chemical species as a function of reaction time, as shown in Figure 2.7. The heights and times of "peaking" of the various chemical species will vary according to the urea/HCHO molar ratio and reaction conditions, but the relative shape of the curves obtained and their relative proportions will remain essentially the same. Taking as 30 min the standard reaction time of a UF resin without a pot-maturing period [1] and fitting the percentages of the various chemical species formed to equation (2.1) will give a total value of adhesion to wood at 30 min reaction time for the UF resin prepared (Table 2.3). At a different reaction time, a different adhesion value will be obtained.

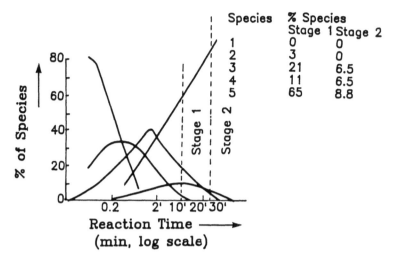

Figure 2.7 Percentages of various UF oligomers in the reaction mix as a function of reaction time for a "one-step" UF resin.

This can be calculated by equation (2.1), the values in Table 2.2, and the equivalent of Fig. 2.7 for any given UF resin. From Fig. 2.7, for the UF resin examined, it is easy to calculate that reaction times longer than 30 min will not show any really significant increases or decreases in adhesion. Shorter reaction times, such as 11 min (Table 2.2 and Fig. 2.7), can give better adhesion with the possible considerable disadvantage, however, that the proportions of the polymeric part of the resin being lower, a weaker resin strength (cohesion) is probable.

After hardening, a certain amount of polymeric material is present and somewhat independent of the degree of condensation (portion of polymer species) before hardening. This still implies that different strengths will be obtained in cured resins derived from differences in the resin while still in liquid form. It is well known that different manufacturing conditions yield resins of very definite differences in strength when cured. Although it thus appears that the monomethylolated and dimethylolated linear species, even in polymeric form, are the ones that contribute most to UF resin adhesion, a balance must always be achieved between the relative proportions of total monomeric and polymeric species unless one wants to maximize adhesion at the expense of total adhesive performance. As a rule of thumb, one should aim not to lower the proportion of the polymeric species below 40 to 50% and the sum of multimethylol ureas plus polymeric species to not much lower

Table 2.3 UF Chemical Species Percentage Distribution According to Formulation and Preparation Process and Corresponding Calculated Comparative Relative Values of Total Adhesion to a Softwood and to Amorphous + Crystalline Cellulose in Wood at 30 min Reaction Time and at Reflux [Equation (2.1)][a].

	UF species distribution (%)			
UF species	Case 1 Fig. 2.7	Case 2 Fig. 2.8	Case 3 Fig. 2.9	Case 4 Fig. 2.10
Urea	0.0	0.0	18.3	12.1
Monomethylolurea	0.0	5.3	12.3	20.2
Dimethylolureas	5.5	14.7	13.0	21.4
Trimethylolurea	4.5	5.3	5.7	5.8
UF polymers (trimer and up)	90.0	74.7	50.7	40.5
Total adhesion per atom of resin mixture to cellulose (kcal mol^{-1}) [equation (2.2)]	0.5543	0.6118	0.6690	0.7390
Total adhesion per atom of resin mixture to a softwood (kcal mol^{-1})	0.4990	0.5647	0.6284	0.6964
Internal bond strength of laboratory particle-board (MPa)	0.37	0.53	0.62	—

[a]Experimental internal bond strength of particleboard prepared with case 1, 2, and 3 resins reacted for only 11 min are shown at the bottom of the table.

than two-thirds of the total chemical species present in the formulation. A certain amount of polymeric material is necessary to avoid too deep a flow of resin to the wood cells and lumina, causing a starved glue line. A certain amount of methylolated species and of free formaldehyde is necessary to guarantee an effective reactivity and short hardening times.

The method exposed lends itself to the optimization of UF resin formulation; for this purpose it is worthwhile to see if (1) a known modification of a UF resin, such as second urea addition, which is well known to improve UF resin adhesion to wood [1,37], will also give adhesion differentiation when using this calculation method; and (2) if the method can be used to predict a UF resin formulation and preparation conditions of even better adhesion and performance.

Second urea addition, the addition toward the end or at the end of UF resin preparation of a certain amount of urea, alternatively expressed as the preparation of UF resin with the total urea content being added partly initially and partly at the end of the reaction, is well known to improve the adhesion to lignocellulosic substrates of a UF resin [1,37]. The process can be depicted as shown in Fig. 2.8, in which a second smaller reaction graph, proportional

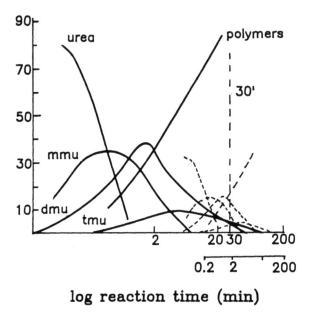

log reaction time (min)

Figure 2.8   Percentages of various UF oligomers in the reaction mix as a function of reaction time for a second-urea-addition UF resin.

to the second urea addition, is superimposed on the original UF reaction. Free formaldehyde is always present and its presence in the original reaction mixture ensures that the second graph dimensions are proportional to the amount of second urea addition, HCHO being available from the existing reaction mixture. The case shown in Fig. 2.8 shows a second urea addition of 30% of total urea. Applying the adhesion values in Table 2.1 to equation (2.1) or (2.2), the results of total adhesion shown in Table 2.3, case 2 are obtained. The adhesion to wood has improved over that of a "straight" UF preparation (Table 2.3, case 1) at the standard reaction time of 30 min, by 13%, not only confirming the experimentally determined effect of the second urea addition but also quantifying it relative to the amount added.

It is of interest, then, to forecast even a better system of UF preparation, to improve adhesion further without detrimental consequences on resin cohesion. Figure 2.9 shows a UF resin in which urea and formaldehyde are added in three equal amounts: initially, at 19 min reaction time and at 30 min reaction time. From equations (2.1) and (2.2) the results shown in Table 2.3, case 3 are obtained: an improvement of adhesion of 26% over a straight UF preparation and of 11% over a second urea preparation. From this reaction

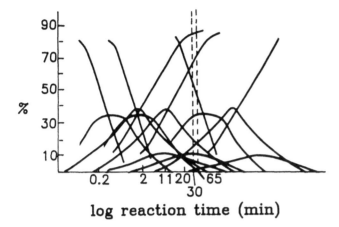

Figure 2.9 Percentages of various UF oligomers in the reaction mix as a function of reaction time for a multiple-successive-addition (three) UF resin.

an additional step to a UF preparation obtained by continuous addition of urea and HCHO during the entire period of reaction warrants investigation. In Fig. 2.10 it can be seen that the various chemical species formed tend, over the reaction period chosen, to achieve a "steady state" as regards their relative proportions. At the standard 30 min reaction time, which generally ensures

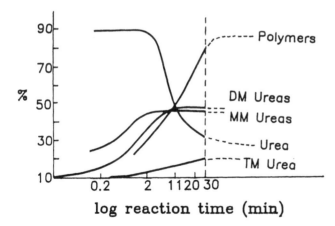

Figure 2.10 Percentages of various UF oligomers in the reaction mix as a function of reaction time showing the reaching of a steady state for a continuous-addition UF resin.

that enough polymeric species are formed to give good resin cohesion, the total adhesion calculated by equation (2.1) or (2.2) is shown in Table 2.3, case 4. The adhesion to wood shows a considerable improvement of 40% over straight UF preparations (Table 2.3, case 1) and of 23% over a second urea addition UF resin (Table 2.3, case 2). There are thus definite indications to switch to continuous addition UF preparations, at least as regards adhesion, as the results demonstrate. By combining reaction curves in other ways, even better resin preparation processes might be obtained, this paragraph reporting just one such case to illustrate the method proposed. A brief comparison of theoretical results with the internal bond strengths of some laboratory-prepared particleboard, prepared by mixing total urea and formaldehyde at the beginning (as in case 1) or using second and third urea additions, is shown in Table 2.3. The trend seen is similar to that observed for the results calculated theoretically. Just these very few tests carried out on particleboard and very few resins tested indicate that there may well be some relation between energy of adhesion and internal bond of particleboard. The number of tests carried out is still too few to gauge if a distinct mathematical relationship exists between the two.

Experimental confirmation of the theoretical approach presented has also been obtained, as regards adhesion, by determining the relative R_f value in paper chromatography of the various UF oligomers. The R_f order is inversely proportional to the energies of adhesion calculated by molecular mechanics at the interface. In Table 2.4 experimental and theoretical results are shown and can be compared. The methyleneurea series—urea, methylenebisurea, and dimethylenetrisurea—appear to respond well to the comparison. Dimethylenetrisurea shows the greatest surface attraction followed by methyl-

Table 2.4 Comparison of Computational Energy Results with Experimental R_f Values for Hydroxymethylureas and Methylene Ureas

Species	Interaction energy (kcal mol^{-1})			
	Average	Minimum	Average per atom	R_f
Monohydroxymethylurea	−14.1	−20.8	−1.18	0.34
N,N'-Dihydroxymethylurea	−17.4	−19.7	−1.09	0.50
Trihydroxymethylurea	−11.2	−12.3	−0.56	0.82
Urea	−4.9	−7.1	−0.61	0.21
Methylenebisurea	−10.3	−12.7	−0.61	0.10
Dimethylenetrisurea (linear)	−16.0	−16.4	−0.61	0.04

enebisurea and then by urea: the same quantitative order in which they appear in the paper chromatogram.

The situation appears to be very different, however, for the series of homologous hydroxymethylureas. In these computational results, the unweighed averages of energy do not even represent the relative R_f values obtained for the three compounds. Thus minimum interaction energy values and average energies divided by the number of atoms in each species are also tabulated against experimental R_f values (Table 2.4). Both of these give good qualitative correspondence with the order in which the hydroxymethylureas appear in the chromatogram. However, only the interaction energy minima for each species also gave good qualitative correspondence with the order in which the compounds within each series, for both series, hydroxymethylureas and methyleneureas, present themselves in the chromatogram.

While within each series there is clear qualitative correspondence between calculated interaction energies and experimental R_f results, this is not the case when comparing the two series with each other. A discontinuity exists: thus it is not possible to reconcile the relative R_f values of any of the methyleneureas to any of the hydroxymethylureas. Thus when just a different chemical group, in this case the hydroxyl of the hydroxymethyl group, is introduced, the computational model used is not yet able to give even qualitatively what the experimental R_f distribution is likely to be. The computational method as described is not able at this stage to place in the correct experimental R_f sequence compounds of different molecular mass that do not belong to the same homologous series, even when the same building blocks, urea and formaldehyde, are used. It appears, then, that other parameters also need to be considered and included in the computational method to model this more complex case. A hypothesis that could be advanced for the two oligomer series considered is that the differences in the extent of solvation known experimentally for these two series of compounds cannot, with a different group present, be modeled exclusively by variations of the dielectric constant of the medium, as done successfully for phenol–formaldehyde adhesives, but might need the introduction of water molecules in the model: a very difficult task.

V. THEORY AND PRACTICE OF LOW-FORMALDEHYDE-EMISSION UF RESINS

The excellent adhesion to lignocellulosics, excellent intrinsic cohesion, ease of handling and application, lack of color in the finished product, and low cost have led the UF resins to be the most widely used adhesives for bonding wood products at present. Their two principal disadvantages—lack of resistance to weather and water, and their susceptibility to emission of formaldehyde vapors—however, also threaten their primary position in the bonding of wood. Both disadvantages are intrinsic properties derived in part from the structural characteristics of the resin and of their type of chemical bond. While the lack of weather resistance is a drawback that can easily be tolerated, as the greatest use of these adhesives is for interior wood products, the UF resin's formaldehyde emission is a drawback that severely affects their applicability in their traditional sector of dominance, the interior wood products market. Confronted with the introduction in many countries of ever more stringent formaldehyde emission requirements, manufacturers have for several years been able to produce UF resins of continually decreasing formaldehyde emission characteristics. At the beginning, decrease in formaldehyde emission from the high values that existed up to the 1970s was easy, centering primarily on a decreased urea/formaldehyde molar ratio. It was also noted that on decreasing the final urea/formaldehyde molar ratio of the resin, the UF adhesive became more difficult to handle, the bond strength was lessened at an equal adhesive content, and consequently, higher adhesive contents were necessary, and sometimes even markedly longer particleboard pressing times were required. The advent of stringent regulations in many countries, particularly in Germany, led to the introduction of UF adhesive resins and systems of even lower emission, capable of providing products in the emission 1 (E1) class—in short, of products in which the formaldehyde equilibrium concentration is 0.1 to 0.06 ppm or lower and with a corresponding perforator test value of 6 to 10 mg of formaldehyde per 100 g of board or less [23,46–49]. Resin manufacturers then began to produce E1 UF adhesives and systems to satisfy both the regulations and the public's increased requirements.

Recent work has indicated that good E1-type UF resins of urea/formaldehyde molar ratio even lower than 1:1.1 can be prepared [25,26,31,32]. The theoretical basis of this finding, useful to advance a tentative theory for low-formaldehyde-emission UF resins, is of interest for the formulation and preparation of UF resins of low formaldehyde emission, first in the laboratory and then at the industrial level. First, these resins can be divided into two broad classes: (1) resins based on the addition of melamine or melamine–formaldehyde (MF) resins to the UF resin, and (2) UF resins in which E1 capability is obtained exclusively by the manipulation of their manufacturing parameters. The former class can also be considered as a subset of the second class.

The underlying principle of an E1 UF resin is that a certain amount of free urea needs to be present to (1) to mop up the greater part of the free formaldehyde that may be present at the end of the preparation, and (2) to mop up the greater part of the free formaldehyde that may be generated during hot curing of the resin. A third possible requirement would be that some free monomeric urea species be left to mop up over a long period of time some of the formaldehyde that may be liberated during the service life of the board.

Such requirements are fundamentally quite divergent and extreme as regards a UF resin. They mean that addition of great amounts of urea are needed, possibly at the end of the reaction; such urea will react with the free HCHO present or generated during hot curing, but will also react with the active methylol groups present on the urea resin itself, severely limiting the possibility of cross-linking of the resin and ultimately diminishing its cured strength. These two sets of divergent requirements indicate that an E1 UF formulation must in general be a compromise between strength and emission requirements. Once this basic conflict of requirements is understood, it can, however, be overcome to attain formulations that give both good strength and low formaldehyde emission.

A UF resin is a mixture of molecular species: namely, methylolureas, UF polymers, and methylolated UF polymers. It has already been proven, both theoretically [25,26,31] and by applied means [32,50], that while monomeric

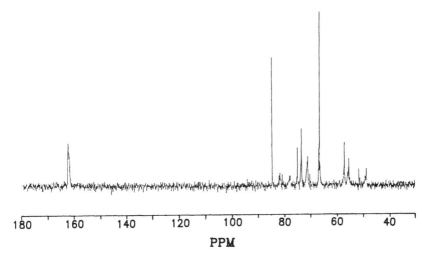

Figure 2.11 ^{13}C NMR spectra of high-formaldehyde-emission resin; note the presence of high amounts of free HCHO (ppm 84.5), absence of free urea, high proportion of methylolated polymers, low proportion of nonmethylolated polymers (ppm 49.0).

and polymeric methylolated species contribute more to the adhesion of the resin to the wood substrate, it is the polymeric fraction (methylolated and nonmethylolated) that contributes most to the cohesion of the resin. Thus a resin to which large amounts of final urea are added will have a proportionally high amount of urea and monomeric methylolated species, giving both good adhesion and low formaldehyde emission, and proportionally, a lower amount of prebuilt polymeric species, giving poor adhesion, hence lower strength. Conversely, a resin of final higher urea/formaldehyde molar ratio, such as the classic UF resins used for the last few decades, will have a large amount of polymeric species and still be heavily methylolated (although the methylolated species will be primarily polymeric) and have a large amount of free and potentially free formaldehyde. These resins will have good cohesion and good adhesion, hence good strength, but very high HCHO emission.

The logical way in which to avoid the conflicting requirements of the two properties wanted is to develop mixtures of species that will give the correct balance of strength and emission for the applications required. There are, of course, various ways to attain the desired E1 effect from particleboard bonded with UF resins [51]:

1. One approach is to decrease the urea/formaldehyde molar ratio to a very low level. This can be achieved easily even by adding further urea to a premanufactured UF resin and possibly reacting for another 5 to 10 min. Although resins are very easy to prepare using this method, the method does present severe limitations. In such resins the last addition of urea is proportionally large, and thus the monomeric species, unreacted urea, and monomethylol and dimethylol ureas are in very high proportions (Fig. 2.12). In such cases a very low urea/formaldehyde molar ratio does ensure a very low formaldehyde emission when the resin glue mix is formed by adding NH_4Cl hardener. At the same time, the high proportion of monomeric species present decreases precipitously both cured strength and bond values. To balance the properties of such a resin correctly is not an easy task. A very simple approach that is also sometimes used is for the particleboard factory to spray a urea solution on the wood particles separately from a non-E1 UF resin. Acceptable results can sometimes be attained with such an unsophisticated system, but considerable experimentation and strict control are often necessary to ensure consistency of production.

2. Another approach is to add melamine as a second or third addition at the end of UF resin preparation. Even small amounts of melamine added in this manner ensure noticeable decreases in the level of formaldehyde emission from the board. The imperviousness of melamine to water ensures that free formaldehyde excesses trapped in methylene melamine bonds are not easily

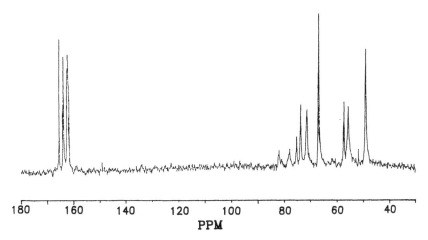

Figure 2.12 ^{13}C NMR spectra of an E1 UF resin; note the presence of very high amounts of methylolated monomers and free urea (ppm 160 to 165), the absence of free HCHO, and the lower proportion of methylolated polymers and much higher proportion of nonmethylolated polymers (ppm 49.0) than in Fig. 2.11.

hydrolyzed. The defects of such an approach are, first, its higher cost, and second, when excessive proportions of final melamine are used, the same defects observed in case 1 above.

3. A better approach is to prepare a UF resin which by itself is not capable of producing E1-type particleboard, but when mixed with additives is easily capable of E1 performance. For this purpose, *accelerators*, simple urea–formaldehyde mixtures of low-condensation prepolymers of urea/formaldehyde molar ratio 1:2 or higher, and *absorbers*, simply urea–formaldehyde mixtures or low-condensation prepolymers of urea/formaldehyde molar ratio 1:0.4 to 0.5 are used. It is then the relative balance of non-E1 UF resin, accelerator, and absorber in the glue mix that determines the strength and emission properties of the board. An advantage of such a system is its flexibility: it adapts to different conditions of application without varying the basic manufacture of the three components. All that changes is the proportion of the components in the glue mix. Thus if higher strength is required, a higher proportion of accelerator is used. If lower emission is required, a higher proportion of absorber is used. The relative proportions of accelerator and absorber added are each generally between 10 and 30% on UF resin solids. This system can, of course, also be used with the resins described in cases 1 and 2.

4. A good method of preparing UF resins of low formaldehyde emission is to increase the number of additions of urea during preparation of the resin. Thus a resin with a third urea addition has better strength and lower emission than one based on a second urea addition procedure. A four-urea-addition resin is better than a three-urea-addition resin, and so on (at equivalent urea/formaldehyde molar ratio percentages). The use of accelerators and absorbers in the glue mix as described in case 3 can also be applied to this type of resin, with the achievement of excellent balance of properties and even finer control in application. Carrying this principle to its limit means that a UF resin prepared by continuous and simultaneous addition of urea and formaldehyde in the reactor during most of the period of manufacture should give excellent results. This is indeed the case, but such a system of manufacture is not easy to control, and small variations in preparation can yield resins quite different in final properties.

It is quite evident, then, that at the molecular level the fundamental principle on which low-formaldehyde UF resins are based is the compromise between the relative proportion, and balance, of monomeric and polymeric species with the highest possible amount of methylolated species on the polymer. This is well illustrated in Figs. 2.11 to 2.13. The resin in Fig. 2.12 is clearly an excellent resin as regards formaldehyde emission, and still likely to present reasonable strength on curing, as observed by the relative proportion

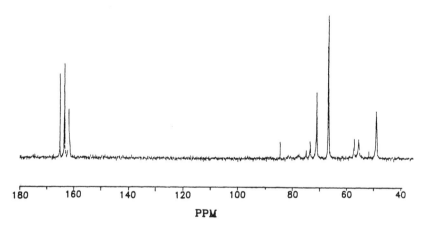

Figure 2.13 ^{13}C NMR spectra of a 50:50 mix of two resins to give an acceptable balance of monomeric methylolated and nonmethylolated species in relation to methylolated and nonmethylolated polymers.

of the region at 160 to 165 ppm in relation to the other regions of the spectrum. The resin shown in Fig. 2.11 is very strong but would have very high emission: it is a classical 1:1.8 molar ratio formulation. The resin in Fig. 2.13 is a hybrid capable of somewhat acceptable strength, of low formaldehyde emission, but not yet capable of E1 performance. It is a resin of type 3 described above, which still needs accelerators and absorbers to give excellent E1 performance.

Physically combining very different resins in some proportion to obtain the desired balance of molecular species and hence the desired balance of cured strength and formaldehyde emission is also an acceptable proposition. Recent results show that at the laboratory level even resins of effective molar ratio 1:0.7 are capable of giving excellent strength and low-E1-class formaldehyde emission results. Comparing industrial and laboratory results, however, it appears that the best balance of properties, or better, the lowest emission with still acceptable board strength, appears to be in the range 1:0.9 to 1:1, particularly not lower than an effective glue mix molar ratio of approximately 1:0.96.

The preparation diagrams of the resins also show interesting features. The third or further urea amount can be added after the maturing time of the second

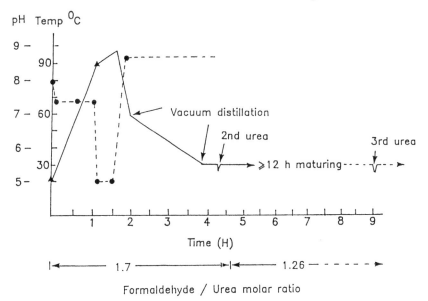

Figure 2.14 Variation of temperature, pH, and formaldehyde/urea molar ratio as a function of reaction time for a UF resin (up to the first maturing period).

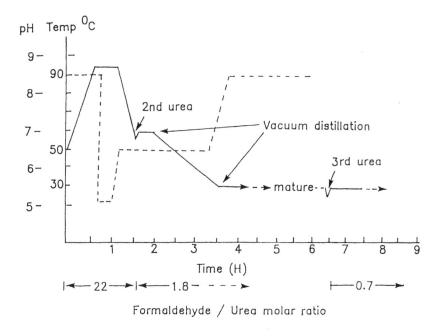

Figure 2.15 Variation of temperature, pH, and formaldehyde/urea molar ratio as a function of reaction time.

urea (Figs. 2.14 and 2.15) or added still in the hot reaction stage of the resin (Fig. 2.16). Figure 2.17 shows, instead, some of the possible variations of process that can be adopted. Thus the second urea amount can be added at 90 to 95°C, at 60°C, or at 30°C. Equally, the third urea addition can be carried out at 90 to 95°C or at 60 to 65°C. It is the balance of reaction time after each urea addition with the reaction mixture pH that will give similar E1 UF resins.

A few other points are of interest. NaCl should always be added to the resin or to the additives, as during board pressing it increases the boiling point of water, giving higher-temperature steam shock in the board core, ensuring higher core temperatures, leading to better formaldehyde reaction and better curing, and contributing to reducing formaldehyde volatility during and after pressing. An interesting point to keep in mind is that often E1-type UF formulations tend to give resins of lower viscosity than those of traditional UF resins. They tend, then, to give a higher proportion of absorption of the resin onto the wood chips. It is advisable in such cases to use a higher solids content of the glue mix, or to prepare the final resin itself at a solids content

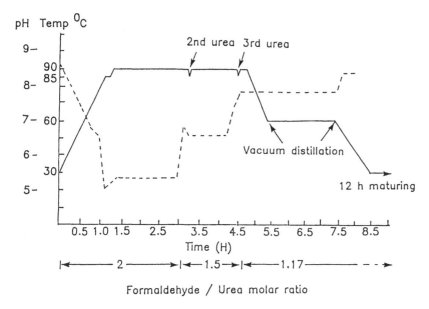

Figure 2.16 Variation of temperature, pH, and formaldehyde/urea molar ratio as a function of reaction time for a UF resin.

higher than the traditional 60 to 65% normally used. This does not, however, solve the problem entirely, but definitely improves strength values.

In conclusion, urea/formaldehyde resins capable of yielding emission 1 (E1) particleboard can be easily produced if a few basic principles are kept in mind. Such formulations need, first, an excess of monomeric non-methylolated and methylolated urea species, which lowers formaldehyde emission and up to a certain extent improves resin adhesion to the wood substrate. Second, they need a certain proportion of nonmethylolated and particularly methylolated polymeric UF species to maintain cross-linking and cohesion, and hence strength of the board, at an acceptable level. A compromise between these two divergent requirements can be achieved by preparing resins that are directly of the E1 type or by a combination of resins and/or urea–formaldehyde mixes each of which has an excess of one of the desired characteristics. At the laboratory level, resins with a urea/formalde-hyde molar ratio as low as 1:0.7 can provide an adequate balance of the two divergent properties sought: strength and low emission. At an industrial level, resins of urea/formaldehyde molar ratio in the range 1:0.9 to 1:1, but

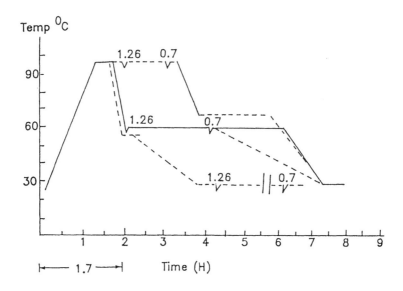

Figure 2.17 Selection of diagrams for temperature and second and third urea additions as a function of reaction time. The lowest segmented line is the diagram of actual preparation. The others are possible alternatives. Decimal numbers in the figure represent the new formaldehyde/urea molar ratio on second and third urea additions.

particularly around 1:0.96, the effective molar ratio of the glue mixes used, appear to give the most appropriate balance of results.

VI. UREA AND ALDEHYDE ALTERNATIVES FOR IMPROVED EMISSION AND WATER RESISTANCE

Recently, the chemistry of UF resins as regards their formaldehyde emission and water repellancy has been approached differently from the methods outlined earlier. One approach [52,53] has been to include long-chain aliphatic primary diamines in the UF resin skeleton. This is of interest because the long alkyl chain connecting the two main groups, once included in the polymer backbone, will improve water resistance and flexibility (by better glue-line stress distribution) in the cured UF resin network [52,53]. The amines were incorporated in the UF polymer chain by three different routes:

1. By direct reaction of the amine with urea and formaldehyde resin synthesis:

    $H_2N–(CH_2)_6–NH_2 + NH_2CONH_2$
    $+ HCHO \rightarrow –NH(–CH_2)_6–NH–CH_2–NHCONH–$

2. By conversion of the primary amine group to its urea derivatives prior to reaction with formaldehyde to form the resin:

    $NH_2–(CH_2)_6–NH_2 + 2NH_2CONH_2$
    $\rightarrow NH_2CONH–(CH_2)_6–NHCONH_2 + 2NH_3$

3. By conversion of the amine to its hydrochloride salt, which is then used instead of ammonium chloride as an acid curing agent:

    $Cl^{-+}NH_2–(CH_2)_6–NH_2^+Cl^-$

The authors [52,53] found that the first modification destroyed resin reactivity and curing. The second and third types of modification gave, instead, resins of various curing reactivities, according to the amine used. These resins presented improved resistance to cyclic stress induced by moisture variations. A clear indication that the flexibility of these resins is an improvement over that of standard UF resins was their reduced tendency to crack and fracture [52,53]. An added positive feature was the decrease in formaldehyde emission induced by decreased density of aminomethylene bonds along the chain of the resins in which the diamines were introduced. The result of particleboard and wood joint cyclic stress testing of UF resins modified by 13% by mass hexamethylene diamine, 16% by mass bis-hexamethylene triamine, or 28% by mass poly(propyleneoxide)triamine showed that such a system has excellent stability even after repeated wet–dry cycles.

It must be pointed out, however, that while proved highly beneficial at the low molar percentage level of addition used, such modifications may become counterproductive if excessive amounts of amine are added. Thus, whereas flexibility and stress distribution are improved noticeably, at the low levels of addition used, a significant increase in the resin of the proportion of each amine over that used by the authors [52,53] might decrease resin cross-linking and increase paracrystallinity to the extent that strength might begin to fall. Of course, such probable disadvantages might well be overcome by increasing the formaldehyde/urea molar ratio to produce the lower formaldehyde emission obtainable with such amine-based modifications. In the future, amine modification might present possibilities even better than the excellent results already achieved and reported [52,53]. A second, but different approach to attaining increased water moisture and decreased formaldehyde emission is to use, in part, an aldehyde different from formaldehyde. Furfural is an option that comes to mind immediately: however, by producing dark boards and being of lower reactivity than formaldehyde, furfural might not be an acceptable alternative, although technically it is an interesting one [60]. The use of aldehydes such as propionaldehyde and butyraldehyde to form urea–alternative aldehyde–formaldehyde resins is possibly a better alternative [54]. The advantage of using these two aldehydes is their water repellancy. Their disadvantage is their reactivity, this being much lower than that of formaldehyde with urea. Urea–propionaldehyde resins have been prepared before [55], more for theoretical and study considerations that with a view to industrial use. Recently, an attempt to use such resins has been reported [54]. The findings were that resins using propionaldehyde exclusively were of such low reactivity that they could not be used for particleboard or other wood-bonding applications [54]. However, resins in which a preformed urea–propionaldehyde resin was reacted with formaldehyde in an alkaline environment to introduce methylol groups on the resin gave a high enough reactivity for laboratory particleboard to be prepared. This resin had lower formaldehyde emission, as expected, and had better water repellancy than that of standard UF resins. The board's boil swelling was very high, but after the first few minutes the swelling did not increase and the board stayed together [54].

VII. COPOLYMER ADHESIVES OF UF WITH DIISOCYANATES

UF resins have been shown to copolymerize readily in water with polymeric MDI [56]. This reaction is based on the reaction of the isocyanate group (–NCO) with the methylol groups on the UF resin, which is much faster than the reaction of –NCO with water. The primary curing mechanism identified is identical to that identified for phenolic resins. Thus

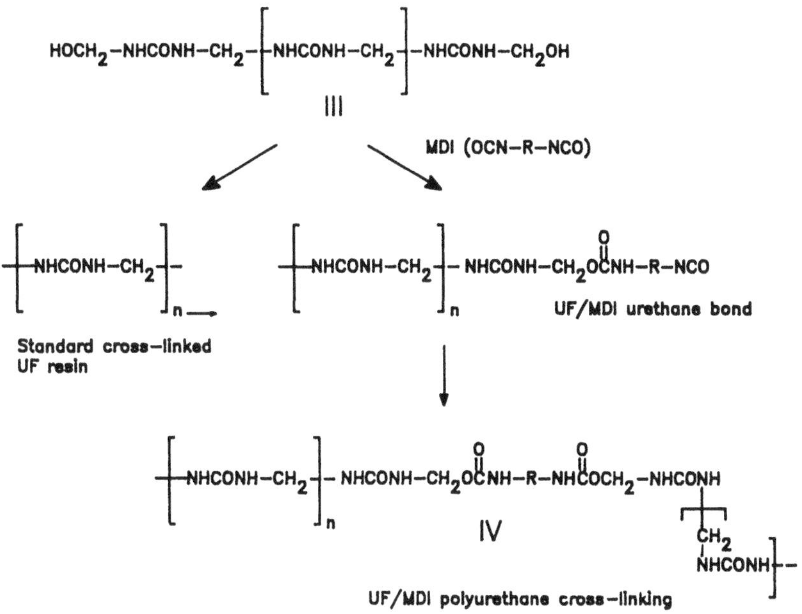

HOCH$_2$−NHCONH−CH$_2$−[−NHCONH−CH$_2$−]−NHCONH−CH$_2$OH

III

MDI (OCN−R−NCO)

[−NHCONH−CH$_2$−]$_n$ ⟶

Standard cross−linked UF resin

[−NHCONH−CH$_2$−]$_n$ − NHCONH−CH$_2$OCNH−R−NCO

UF/MDI urethane bond

[−NHCONH−CH$_2$−]$_n$ − NHCONH−CH$_2$OCNH−R−NHCOCH$_2$−NHCONH

IV

CH$_2$
NHCONH−−

UF/MDI polyurethane cross−linking

The mixture of watery UF resin and MDI goes through a rapidly formed prepolymer stage as soon as the two resins are mixed at ambient temperature. Such a mix presents an adequately long pot life to render it suitable for industrial use. The normal amount of ammonium chloride is generally used as a hardener of the UF portion of the resin [57]. The adhesive system was found to have a good application life even when applied to plywood veneers before pressing. The results on veneers of laboratory plywood panels from species particularly refractory to bonding show (Tables 2.5 and 2.6) that a variety of performances can be attained by varying the relative proportions of UF resin solids and of polymeric MDI [57]. Such results indicate that performance comparable or better than that attainable with good melamine or phenolic resins is attainable at a very competitive cost.

Such a system, which is excellent for plywood, might not be readily usable for particleboard, due to its higher viscosity range. As regard plywood application, the system has the unusual characteristic of flowing and spreading so well as to give the illusion of being of much lower apparent viscosity. In this context, of particular interest is the application of UF resins of U/F molar ratio lower than 1 in combination with diisocyanates to prepare low-formaldehyde-emission boards of good strength, which has been proposed as an alternative technology for the production of interior E1 particleboard [58,59].

Table 2.5 Knife Test Results According to British Standard BS 1088–1957 of Laboratory-Prepared Plywood Panels Using MDI–UF on Tepa (*Laurelia philippiana*) [57] Hardwood Veneers

	Knife test rating		
MDI/UF solids ratio	Dry	24-h soak	6-h boil
50 : 50	10	9	5
40 : 60	9	9	5
30 : 70	8	4	0
20 : 80	9	3	1
Best commercial PF control	4	2	0
Commerical MF control	6	3	0
Commerical UF control	6	0	0

VIII. FINE TUNING OF HARDENERS FOR UF RESINS

The range of hardening times that can be achieved with various hardeners is considerable and sufficient to satisfy the most diverse application requirements. It must be kept in mind that UF resin hardeners are really only accelerators of UF resin curing, introducing a catalytic or, rather, an activating effect only. They do not, in the main, determine the final strength of the cured resin but only the rate at which this final strength will be achieved. Options exist to vary ambient-temperature hardening times, hence pot life, of UF resins: (1) by changing the proportion of hardener used, (2) by changing the

Table 2.6 Representative Glue Mixes for MDI–UF Laboratory Resins in Table 2.5

	MDI/UF 40 : 60 parts by mass
UF resin, 60% solids (U/F = 1 : 1.3)	130
MDI (Bayer 44V20)	52
NH_4Cl (20% water solution)	10
Wood flour, 230 mesh	2

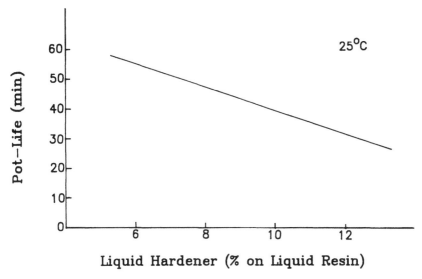

Figure 2.18   Pot-life of UF glue mix as a function of % liquid hardener added (hardener = 6.5:1.5 mass/mass ammonium chloride/phosphoric acid 8% water solution).

relative proportion of chemicals in the hardener, or (3) by changing the composition of the hardener.

The first option is somewhat limited, as variations in only a limited range can be obtained. In Fig. 2.18 is shown the variation in pot life at ambient temperature of an ammonium chloride/phosphoric acid 6.5:1.5 mass/mass 8% water solution on a UF resin. It is clear that increasing the percentage hardener from 6% of liquid UF resin to 12% shortens the ambient temperature pot life by only about 35%.

The second approach is to change the relative proportions of the same chemicals in the hardener. Use of the type of fast cold-set hardener shown in Figure 2.18 yields different pot lives for the UF resin when the relative proportions of ammonium chloride and phosphoric acid are changed. Table 2.7 gives a clear example: by decreasing the amount of phosphoric acid, the pot life is lengthened, and vice versa. Variations of pot life through use of this second approach are much more noticeable for slow hardeners than for fast ones (Table 2.8).

When greater variations in ambient-temperature setting times and pot lives are required, the composition of the hardener must be changed. Thus a 15% solution of ammonium chloride in water will give pot lives for 11% mass addition on a liquid UF resin syrup of approximately 5 h. A 4% ammonium

Table 2.7 Effect on Hardening Time of Changing the Relative Proportion of Chemicals in a Hardener

	Hardener				
	1	2	3	4	5
Water	92	92.5	93	90	88
Ammonium chloride	6.5	6.5	6.5	8.5	10.5
Phosphoric acid	1.5	1.0	0.5	1.5	1.5
Pot life (min)	40	47	66	39	35

chloride plus 6% hexamethylenetetramine water solution used as hardener will give a pot life of 7 to 9 h.

IX. FORMULAS

A. Resin 1 (Non-E1)

To 1000 parts by mass of 42% formaldehyde (not containing methanol) solution are added a 22% NaOH solution to pH 8.3 to 8.5 and 437 parts by mass of 99% urea, and the temperature is raised in ±50 min from ambient to 90°C while maintaining a pH range of 7.3 to 7.6 by small additions of 22% NaOH and a temperature of 90 to 91°C for another 20 min until the turbidity point is reached. The pH is corrected to 4.7 to 5.1 by addition of 30% formic acid and the temperature raised to 98°C. The water tolerance

Table 2.8 Effect on Hardening Time of Changing Hardener Composition for a Slow Hardener

Hardener:	1	2	3	4	5
Water (g)	9.15	8.95	8.75	9.1	8.8
Citric acid (g)	0.2	0.4	0.6	0.4	0.4
Diammonium phosphate (g)	0.65	0.65	0.65	0.5	0.8
Glue mix:	1	2	3	4	5
UF 65% syrup (g)	25	25	25	25	25
Hardener (g)	2.5	2.5	2.5	2.5	2.5
Gel time	3 h 29 min	2 h 47 min	2 h 12 min	2 h 48 min	2 h 44 min

point is reached in ±18 min and then the pH adjusted to 8.7. Vacuum distillation of reaction water with concomitant cooling is then initiated and terminated 100 min later. The resin is cooled to 40°C and 169 parts by mass of second urea is added, the pH adjusted to 8.5 to 8.7, and the resin allowed to mature at 30°C for 24 to 48 h. The resin solids content is 60%, density 1.268 g cm^{-3}, free HCHO 0.40%, viscosity 200 cP, and pH 8.

B. Resin 2 (E1)

To 1000 parts by mass of resin 1 above is added 367 parts by mass of third urea. The mix is heated to 65°C in 45 min and maintained at 65°C for 5 to 10 min, then cooled and stored. The resin has 70% solids, the urea/HCHO molar ratio is 1:0.7, the free HCHO content 0.11%, density 1.288 g cm^{-3}, pH 7.8, viscosity 202 cP, and gel time 68 s.

C. Resin 3 (E1)

Same as resin 2 above, but 413 parts by mass of third urea is added.

D. Resin 4 (Non-E1 when Alone)

To 1000 parts by mass of 33% formaldehyde solution are added 22% NaOH to pH 8.3 to 8.5 and 390 parts by mass of 99% urea solids; the temperature is raised in ±50 min from ambient to 85°C while maintaining the pH at 6.5 to 7.0. The pH is corrected to 4.9 to 5.2 by a single addition of 30% formic acid and the temperature raised to 90 to 92°C and maintained at ±90°C for 140 min while maintaining the pH at approximately 5.2 to 5.3. The pH is then adjusted to 6.5 to 6.8, 130 parts by mass of second urea solids added, and the temperature maintained at ±90°C for another 75 min. The pH is then adjusted to 7.5 to 7.6 and 145 parts by mass of third urea solids added. The temperature is maintained as closely as possible at ±90°C for 10 min after third urea addition and the pH adjusted to 7.7 to 8.0. The resin is then cooled to 60°C in ±30 min, and when 60°C is reached vacuum distillation of reaction water is started and continued for approximately 2 h. The resin, after distillation, is cooled to 30°C in ±60 min while maintaining the pH at ±8.3, and matured for a minimum of 12 h in storage. The resin has a 66% solids, a urea/HCHO molar ratio of 1:1.17, pH 8, density 1.290 g cm^{-3}, viscosity 320 cP, free HCHO 0.23%, and gel time 63 s.

E. Resin 5 (E1)

To 1110 parts by mass of water are added 1000 parts by mass of 97% paraformaldehyde powder and 30% NaOH to a pH of 8.5 to 9.5. The mixture is heated to 50°C, 880 parts by mass urea solids added, and the temperature raised in ±45 min to 95°C and maintained at 95°C for 15 min while maintaining the pH at 8.0 to 8.5. Thirty percent formic acid is then added to a pH of 5.1 to 5.2 and the mixture reacted for 20 min at 95°C. The pH is then adjusted to 6.5 to 6.6, the resin cooled to 60°C, and 195 parts by mass second urea added and reacted at 60°C for 20 min. The resin is then cooled to room temperature in ±45 min, the pH adjusted to 7.8 to 8.4, and the resin left to mature in storage for 48 h. Then 1340 parts by mass third urea is added and the resin heated to 65°C in ±45 min, maintained at 65°C for 10 min, and then cooled to room temperatures in 30 to 40 min. The resin solids content is 64%, pH 8, viscosity 79 cP, and free HCHO 0.16%.

F. Resin Accelerator (Non-E1 when Alone)

To 780 parts by mass water are added 1000 parts by mass of 42% formaldehyde solution and 500 parts by mass NaCl, followed by stirring at 20 to 25°C until the NaCl is dissolved. To the mixture is added 400 parts by mass urea and the pH adjusted to 8.7 to 8.9 with 22% NaOH solution. The temperature is maintained at between 25 and 30°C. The solids content is 48%.

G. Resin HCHO Emission Suppressant (Urea/HCHO = 1.0.4)

To 830 parts by mass water are added 1000 parts by mass of 42% formaldehyde solution and 215 parts by mass of NaCl under mechanical stirring until dissolved. Then are added triethanolamine 1.8, borax 2.2, and hexamethylenetetramine 11 parts by mass followed by 20% NaCl solution 44 parts and 33% NaOH 14 parts. Then 1100 parts urea is added, followed when dissolved first by 65 parts of 25% NH_3 and by another 1000 parts urea, all under mechanical stirring. Gentle warming is used during the entire preparation to maintain a temperature of 25 to 30°C. The pH is adjusted to ±9.5 with 33% NaOH solution.

REFERENCES

1. A. Pizzi, ed., *Wood Adhesives: Chemistry and Technology*, Vol. 1, Marcel Dekker, New York, 1983.
2. G. Zigeuner, *Fette Seifen Austrichm.*, *56*: 973 (1954); *57*: 14, 100 (1955).
3. L. E. Smythe, *J. Phys. Colloid Chem.*, *51*: 369 (1947).
4. G. A. Growe and C. C. Lynch, *J. Am. Chem. Soc.*, *70*: 3795 (1948); *71*: 3731 (1949); *75*: 574 (1953).

5. L. Bettelheim and J. Cedwall, *Sven. Kem. Tidskr.*, *60*: 208 (1948).
6. G. Smets and A. Borzee, *J. Polym. Sci.*, *8*: 371 (1952).
7. J. I. de Jong and J. de Jonge, *Trav. Chim. Pays-Bas*, *72*: 207 (1953).
8. J. I. de Jong and J. de Jonge, *Trav. Chim. Pays-Bas*, *72*: 213 (1953).
9. J. Billiani, K. Lederer, and M. Dunky, *Angew. Makromol. Chem.*, *180*: 199 (1990).
10. M. Dunky and K. Lederer, *Angew. Makromol. Chem.*, *102*: 199 (1982).
11. F. Kasbauer, D. Merkel, and O. Wittmann, *Z. Anal. Chem.*, *281*: 17 (1976).
12. M. Dunky, *Holzforsch. Holzverwert.*, *37*(4): 75 (1985).
13. V. Horn, G. Benndorf, and K.-P. Rädler, *Plaste Kautsch.*, *25*: 570 (1978).
14. M. Chiavarini, N. Del Fanti, and R. Bigatto, *Angew. Makromol. Chem.*, *46*: 151 (1975).
15. B. Tomita and S. Hatono, *J. Polym. Sci. Polym. Chem. Ed.*, *16*: 2509 (1978).
16. J. R. Ebdon and P. E. Heaton, *Polymer*, *18*: 971 (1977).
17. R. Taylor, R. J. Pragnell, J. V. McLaren, and C. E. Snape, *Talanta*, *29*: 489 (1982).
18. R. M. Rammon, W. E. Johns, J. Magnuson, and A. K. Dunker, *J. Adhes.*, *19*: 115 (1986).
19. G. Maciel, N. Szeverenyi, T. Carly, and G. Meyers, *Macromolecules*, *16*: 598 (1983).
20. E. E. Ferg, M.Sc. thesis, University of the Witwatersrand, Johannesburg, South Africa, 1993.
21. E. E. Ferg, A. Pizzi, and D. C. Levendis, *J. Appl. Polym. Sci.*, *50*: 907 (1993); *Holzforsch. Holzverwert.*, in press (1993).
22. T. Matsumoto, *Rinzyo Shikenjo Kenkya Hokoku*, *262*: 41 (1976).
23. R. Marutzky, Chapter 10 in *Wood Adhesives: Chemistry and Technology*, Vol. 2 (A. Pizzi, ed.), Marcel Dekker, New York, 1989.
24. E. Roffael and L. Melhorn, *Holz-Zentralbl.*, *102*: 2202 (1976).
25. D. Levendis, A. Pizzi, and E. Ferg, *Holzforschung*, *46*: 263 (1992).
26. A. Pizzi, *J. Adhes. Sci. Technol.*, *4*(7): 573 (1990).
27. W. E. Johns and W. K. Motter, *Proceedings of the 1988 Particleboard Symposium*, Washington State University, Pullmann, Wash., 1988, pp. 151–158.
28. A. K. Dunker, W. E. Johns, R. Rammon, B. Farmer, and S. Johns, *J. Adhes.*, *19*: 153 (1986).
29. T. J. Pratt, W. E. Johns, R. M. Rammon, and W. L. Plagemann, *J. Adhes.*, *17*: 275 (1985).
30. A. Pizzi and N. J. Eaton, *J. Adhes. Sci. Technol.*, *1*: 191 (1987); 7: 81 (1993); *Chem. Phys.*, *164*: 203 (1992).
31. A. Pizzi, *J. Adhes. Sci. Technol.*, *4*(7): 589 (1990).
32. A. Pizzi, *Holzforsch. Holzverwert.*, *43*(3): 63 (1991).
33. J. I. de Jong, J. de Jong, and H. A. K. Eden, *Trav. Chim. Pays-Bas*, *72*: 88 (1953); *73*: 118 (1954).
34. Y. Nakarai, *Wood Ind. Jpn.*, *3*: 19 (1952).

35. M. Inoue and M. Kawai, *Res. Rep. Nagoya Munic. Ind. Res. Inst.*, *12*: 73 (1954).
36. K. Horioka, M. Noguchi, and M. Saito, *Bull. Gov. For. Exp. Stn. Tokyo*, *113*: 2 (1959).
37. P. R. Steiner, *For. Prod. J.*, *23*: 32 (1973).
38. R. Steele and L. E. Giddens, *Ind. Eng. Chem.*, *48*: 110 (1956).
39. B. A. Kottes Andrews, R. M. Reinhardt, J. G. Frick, and N. R. Bertoniere, Chapter 5 in *Formaldehyde Release from Wood Products* (B. Meyer, B. A. Kottes Andrews, and R. M. Reinhardt, eds.), ACS Symposium Series No. 316, American Chemical Society, Washington, D.C., 1986.
40. W. E. Johns, Chapter 3 in *Wood Adhesives: Chemistry and Technology*, Vol. 1, (A. Pizzi, ed.), Marcel Dekker, New York, 1983.
41. A. Pizzi, N. J. Eaton, and M. Bariska, *Wood Sci. Technol.*, *21*: 235 (1987).
42. A. Pizzi, M. Bariska, and N. J. Eaton, *Wood Sci. Technol.*, *21*: 317 (1987).
43. A. Pizzi and N. J. Eaton, *J. Macromol. Sci. Chem. Ed.*, *A22*: 105 (1985).
44. C. Skaar, Chapter 3 in *The Chemistry of Solid Wood* (R. Rowell, ed.), American Chemical Society, Washington, D.C., 1984.
45. A. Pizzi and N. J. Eaton, *J. Macromol. Sci. Chem. Ed.*, *A24*: 1065 (1987).
46. H.-J. Deppe, in *Luftqualitat in Innenraum* (K. Aurand, B. Seifert, and J. Wegner, eds.), Fisher Verlag, Stuttgart, 1982, pp. 91–128.
47. H.-J. Deppe, *Holz-Kunstoffverarb.*, *20*(7/8): 12 (1985).
48. H.-J. Deppe, *Holz-Kunstaffverarb.*, *21*(7/8): 12 (1986).
49. R. Marutzky, *Adhaesion*, *28*(12): 10 (1984).
50. S. Chow and P. R. Steiner, *For. Prod. J.*, *23*(12): 32 (1973).
51. A. Pizzi, L. Lipschitz, and J. Valenzuela, *Holzforschung*, *48*(3) (1994).
52. R. O. Ebewele, G. E. Myers, B. H. River, and J. A. Koutsky, *J. Appl. Polym. Sci.*, *42*: 2997 (1991).
53. R. O. Ebewele, B. H. River, G. E. Myers, and J. A. Koutsky, *J. Appl. Polym. Sci.*, *43*: 1483 (1991).
54. U. Duvenhage, M.Sc. thesis, University of the Witwatersrand, Johannesburg, South Africa, 1993.
55. K. K. Ibragimov and K. R. Rustamov, *Zh. Fiz. Khim*, *44*(6): 1563 (1970).
56. A. Pizzi and T. Walton, *Holzforschung*, *46*(6): 541 (1992).
57. A. Pizzi, J. Valenzuela, and C. Westermeyer, *Holzforschung*, *47*(1): 68 (1993).
58. A. Lambuth, private communication, 1993.
59. Netherland Patent.
60. A. Pizzi, Holz Roh Wezkstoff, Kurz Originalia , 48 (1990).

3

Melamine–Formaldehyde Adhesives

I. INTRODUCTION

Melamine–formaldehyde (MF) and melamine–urea–formaldehyde (MUF) resins are among the most used adhesives for exterior and semiexterior wood panels and for the preparation and bonding of both low- and high-pressure paper laminates and overlays. Their much higher resistance to water attack is their main distinguishing characteristic from UF resins. MF adhesives are expensive. For this reason MUF resins which have been cheapened by addition of urea are often used. Notwithstanding their widespread use and economical importance, the literature on melamine resins is only a small fraction of that dedicated to UF resins. Often, MFs and MUFs are described in the literature only as a subset of UF amino resins. This is not really the case, as they have peculiar characteristics and properties all of their own, which are in certain respects very different from those of UF adhesives.

A. Uses

MF resins are used as adhesives for exterior- and semiexterior-grade plywood and particleboard. In this application their handling is very similar to that of UF resins for the same use, with the added advantage of excellent water and weather resistance. MF resins are also used for the impregnation of paper sheets in the production of self-adhesive overlays for the surface of wood-based panel products and of self-adhesive laminates. In this application the impregnation substrate, α-cellulose paper, is impregnated thoroughly by immersing it in the resin solution, squeezing it between rollers, and drying without curing it to proper flow by passing it through an air-draft tunnel oven at 70 to 120° C at ± 10m s^{-1}. The dry MF-impregnated sheets can then be bonded by one of two main processes:

1. Sheets of MF-impregnated paper, consisting of one surface layer or a few surface layers, are bonded together with a substrate of paper sheets impregnated with phenolic resins to form laminates of variable thickness. In the MF-impregnated paper is the dry but still active MF resin, which functions as an adhesive binding MF-impregnated sheets to MF-impregnated sheets and also at the interface between MF- and PF-impregnated layers. These laminates are high-pressure laminates.

2. The MF in an impregnated paper sheet is not completely cured but still has a certain amount of residual activity and is applied directly in a hot press, in a single sheet, on a wood-based panel to which it bonds by completing the MF adhesive curing process.

Press platens are made from stainless steel– or chromium-plated brass and copper. The chromium layer preserves the surface quality longer than does ordinary steel. The MF laminates exhibit a remarkable set of characteristics. Because of their unusual chemical inertness, nonporosity, and nonabsorbance, they resist most substances, such as mild alkalies and acids, alcohols, solvents such as benzene, mineral spirits, natural oils, and greases. No stains are produced on MF surfaces by these substances. This remarkable resistance, in addition to almost unlimited coloring and decorating possibilities, has resulted in the extensive use of MF-laminated wood-based panel products for tabletops, sales counters, laboratory benches, heavy-duty work areas in factories and homes, wall paneling, and so on.

B. Condensation Reactions

The condensation reaction of melamine (I) with formaldehyde (Fig. 3.1) is similar to but different from the reaction of formaldehyde with urea. As with urea, formaldehyde first attacks the amino groups of melamine, forming methylol compounds. However, formaldehyde addition to melamine occurs more easily and completely than addition to urea. The amino group in melamine easily accepts up to two molecules of formaldehyde. Thus complete methylolation of melamine is possible, which is not the case with urea [1].

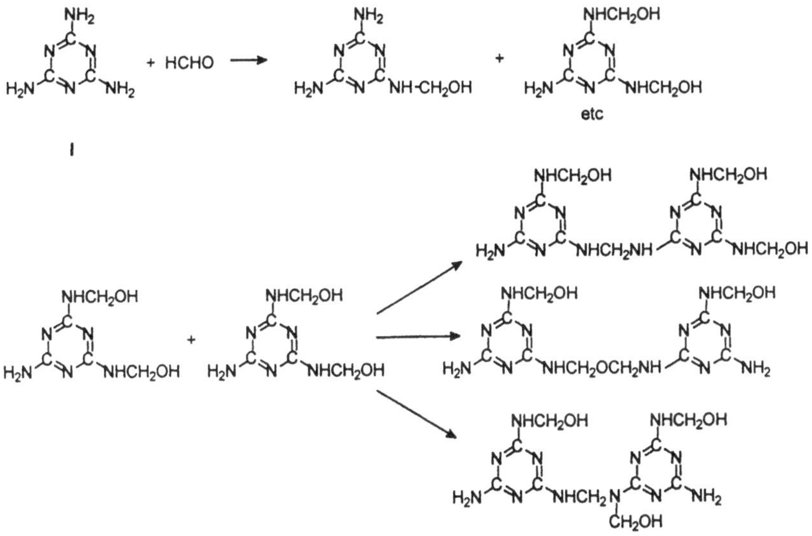

Figure 3.1 Methylolation (hydroxymethylation) and subsequent condensation reactions to form MF adhesive resins.

Up to six molecules of formaldehyde are attached to a molecule of melamine. The methylolation step then leads to a series of methylol compounds with two to six methylol groups. Because melamine is less soluble than urea in water, the hydrophilic stage proceeds more rapidly in MF resin formation. Therefore, hydrophobic intermediates of the MF condensation appear early in the reaction. Another important difference is that MF condensation to give resins, and their curing, can occur not only under acid conditions, but also under neutral or even slightly alkaline conditions. The mechanism of the further reaction of methylol melamines to form hydrophobic intermediates is the same as for UF resins, with splitting off of water and formaldehyde. Methylene and ether bridges are formed and the molecular size of the resin increases rapidly. These intermediate condensation products constitute the bulk of the commercial MF resins. The final curing process transforms the intermediates to the desired MF insoluble and infusible resins through the reaction of amino and methylol groups, which are still available for reaction.

A simplified schematic formula of cured MF resins has been given by Koehler [2] and Frey [3]. They emphasize the presence of many ether bridges besides unreacted methylol groups and methylene bridges. This is because in curing MF resins at temperatures of up to 100°C, no substantial amounts of formaldehyde are liberated. Only small quantities are liberated during curing up to 150°C. However, UF resins curing under the same conditions liberate a great deal of formaldehyde.

At the condensation stage attention must be paid to the formation of hydrolysis products of the melamine before preparation starts. The hydrolysis products of melamine are obtained when the amino groups of melamine are gradually replaced by hydroxyl groups. Complete hydrolysis produces cyanuric acid.

melamine   ammeline

ammelide   cyanuric acid

Ammeline and ammelide can be regarded as partial amides of cyanuric acid. They are acid and have no use in resin production. They are very undesirable by-products of the manufacture of melamine because of their catalytic effect in the subsequent MF resin production, due to their acidic nature. If present, both must be removed from crude melamine by an alkali wash and/or crystallization of the crude melamine.

C. Mechanisms and Kinetics

The mechanism of the initial stages of the reaction of melamine with formaldehyde, leading to the formation of methylol melamines, is very similar to that of urea. The reaction mechanism of the acid-catalyzed condensation reactions of methylol melamines to form polymers and resins has been elucidated by Sato and Naito [4]. Melamine and formaldehyde react in a similar way to urea and formaldehyde, although basic differences are evident in the reaction rates and mechanism. The primary products of reaction are methylol melamines, and evidence indicates that such compounds are formed only at ambient or higher temperature, except in acid pH ranges. The reaction is reversible throughout the pH range. Its forward rate is proportional to either [melamine][HCHO], or [melamine][H^+CHOH], or [melamine$^+$][HCHO], according to the pH used.

Methylol melamine forms dimers by condensation with melamine under neutral and acid conditions (70° C). This process is irreversible. The initial hydroxymethylation is very rapid. Its rate is determined by the condensation of conjugated acids of methylol melamines with melamine. The reaction rate is proportional to that of [melamine]^2[HCHO] [5]. Where the [mineral acid]/[melamine] ratio is 0.0:1.0, the early-stage hydroxymethylation of melamine is dependent on the concentration of the melamine molecule (base species) MH and its conjugated acid MH_2^+ in the following manner [6]:

$$\text{rate} = k_{H_2O}[MH][HCHO] + k_H [MH_2^+][HCHO]$$
$$+ k_{MH_2^+}[MH_2^+] [MH] [HCHO] + k_{MH}[MH]^2 [HCHO]$$

In the absence of added acid—thus when the ratio [mineral acid]/[melamine] is zero—the rate of the reaction can be represented as

$$\text{rate} = k_{H_2O}[MH] [HCHO] + k_{MH}[MH]^2 [HCHO]$$

The condensation reaction has been studied by investigating the kinetics of the initial state of the condensation of di- and trimethylol melamine (MF_2 and MF_3) in the pH range 1 to 9. Regardless of pH, the initial rate is equal to [4]

$$\text{rate} = k[MF_n]^2 \qquad (n = 2 \text{ or } 3)$$

In the presence of mineral acid, the main reaction at the early stage of the condensation is the reaction between the methylol melamine molecule and its conjugated acid (MF_nH^+) [7]. This was found at an [acid]/[MF_n] ($n = 2$ or 3) ratio lower than 1.0 (pH 2.7). With an [acid]/[MF_n] ratio higher than 1.0 to 1.2 (pH < 2), the main condensation takes place between the conjugated acids themselves.

At equal pH values the condensation rate of trimethyol melamine is considerably faster than that of dimethylol melamine. This is the opposite of the rates of mono- and dimethylolurea. This means that while the nitrogen of the amido group in the case of urea is more reactive, and therefore more nucleophilic, than the nitrogen of the amidomethylol group, the opposite is true in the case of melamine. The reaction of MF_2 is mainly between the carbon of the methylol group next to the nitrogen in HM^+CH_2OH, and the nitrogen of the amino group in MCH_2OH.

With regard to MF_3, the condensation is mainly between the carbon of the methylol group next to the charged nitrogen in H^+MCH_2OH and the nitrogen of the aminomethylol group in MCH_2OH [4]. The condensation rate therefore increases with the increasing electrophilicity of the carbon of the methylol group and with increasing nucleophilicity of the nitrogen of the amino group or aminomethylol group. Therefore, in MF_3 the carbon in HM^+CH_2OH is more electrophilic than the same carbon in MF_2. On the other hand, the nitrogen of the aminomethylol group in HM^+CH_2OH of MF_3 is less nucleophilic, and therefore less reactive, than the nitrogen of the amino group of MF_2. The effects of the carbon and nitrogen atoms are consequently opposite to each other in the MF_n condensation. Since the effect of the carbon is greater than the effect of the nitrogen on the reaction rate, the MF_3 condenses faster than MF_2. At lower pH values the effect of the nitrogen becomes negligible and MF_3 is even faster than MF_2 in condensing to polymers.

The difference between the kinetic behavior of urea and melamine can be ascribed to the different effect of the nitrogen atom in the two compounds. With regard to the formation of methylol compounds as a result of hydroxymethylation, the functionality of melamine has been observed to be 6 against formaldehyde. Similarly, melamine easily reacts with formaldehyde to form MF_3. It also forms MF_6 in concentrated formaldehyde. Urea, for instance, readily forms dimethyolurea, but forms trimethylolurea with marked difficulty and never forms tetramethylolurea. These results suggest that the nitrogen of the amidomethylol group in methylolurea is considerably less nucleophilic than the nitrogen of the amido group in urea. However, the nitrogen of the aminomethylol group in methylol melamine is not markedly less nucleophilic than the nitrogen of the amino group in melamine. Presumably, this is due to the difference in basicity between urea and melamine. The same is true of their condensation reactions.

D. Application

MF resins produce high-quality plywood and particleboard because their adhesive joints are boilproof [8,9]. Considerable discussion has occurred and many investigations have been carried out on the weather resistance of MF adhesives. Many authors uphold the good weather resistance of the more recently developed MF adhesives, especially those in which small amounts of phenol have been incorporated. The more general trend, however, is to consider the wood products manufactured with these resins as capable of resistance to limited weather and water exposure only, such as in flooring applications, rather than being capable of true exterior-grade weather resistance, for which phenolic adhesives are preferred. In Fig. 3.2 is shown the different behavior of MF- and PF-bonded particleboard to a series of wet–dry cycles. Whereas PF-bonded boards initially deteriorate rapidly, then stabilize to a constant swelling value, the MF bonded particleboards have slower initial deterioration but never stabilize and continue to deteriorate with time and further wet–dry cycles. This indicates that MF-bonded wood products are not completely impervious to further water attack, indicating the fundamental susceptibility of the aminoplastic bond to water. The rate of deterioration, and therefore bond hydrolysis, is faster as temperature rises. Considering the insolubility of melamine in cold water, this is quite understandable.

II. RESIN PREPARATION

A. MF Resins

Preparation of MF resins does not present undue difficulty. Industrial equipment is standard: a stainless steel or glass-lined reactor equipped with suitable cooling and heating coils or jacket, good mechanical stirring, and reflux and vacuum distillation condensers for dewatering the reaction mixture toward the end of resin manufacture. Formulations that do not need dewatering are also used commercially. The molar ratio of melamine to formaldehyde is

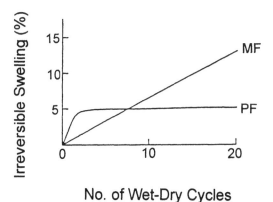

Figure 3.2 Irreversible thickness swelling characteristics of particleboard bonded with MF and PF adhesives during a wet–dry cycle.

generally 1.5 to 2.5. Lower formaldehyde ratios can be used, depending on the application for which the resin is needed and on the system of manufacture.

Due to their characteristic rigidity and brittleness in their cured state, when MF resins are used for impregnated paper overlays, small amounts, typically 3 to 5%, of modifying compounds are often copolymerized with the MF resin during its preparation to give the finished product better flexibility. Most

acetoguonamine

E-caprolactame

p-toluene sulphonamide

commonly used are acetoguanamine, ε-caprolactam, and p-toluenesulfonamide.

The effect of these is to decrease cross-linking density in the cured resin, due to the lower number of amidic or aminic groups in their molecules. Thus, in resin segments where they are included, only linear segments are possible, decreasing the rigidity and brittleness of the resin. Acetoguanamine is most used for modification of resins for high-pressure paper laminates, while caprolactam, which in water is subject to the following equilibrium, is used primarily for low-pressure overlays for particleboard. Small amounts of noncopolymerized plasticizers such as diethylene glycol can be used for the same purpose. Due to the peculiar structure of the wood product itself, MF adhesives for particleboard generally do not need the addition of these modifiers. Often, a small amount of dimethylformamide, a good solvent for melamine, is added at the beginning of the reaction to ensure that all the melamine is dissolved and is available for reaction. Sugar is often added to cheapen the resin. The aldehyde group of sugars have been proven capable of condensing with the amide groups of melamine and hence copolymerizing in the resin. Their quantity in MF resins must be limited to very low percentages, and if possible, sugars should not be used at all, as they tend to cause aging, yellowing, crazing, and cracking of cured MF paper laminates and to badly affect the adhesive's long-term water resistance in both plywood and particleboard.

$$\underset{\substack{\text{H}_2\text{C}-\text{CH}_2 \\ \text{H}_2\text{C} \qquad \text{CH}_2 \\ \text{H}_2\text{C} \quad \text{C}=\text{O} \\ \text{N} \\ \text{H}}}{} \rightleftharpoons H_2N-(CH_2)_5-COOH$$

MF adhesive resins for plywood and particleboard must be prepared to quite different characteristics than those for paper impregnation. The latter must have lower viscosity but still high resin solids content because they need to penetrate the paper substrate comprehensively to a high resin load, be dried without losing adhesive capability, and only later be able to bond strongly to a substrate. MF adhesive resins for plywood and particleboard are, instead, generally more condensed, to obtain lower penetrability of the wood substrate (otherwise, some of the adhesive is lost by overpenetration into the substrate). The reverse applies for paper substrates, where the contrasting characteristics wanted (both good paper penetration and fast curing) can be obtained in several ways during resin preparation. These characteristics can be achieved by producing, for instance, a resin of a lower degree of condensation and a high methylol group content. Typically, an MF resin of a lower level of condensation and a melamine/formaldehyde molar ratio of 1:1.5 to 2 will give the characteristics desired. Its high methylol content and somewhat lower degree of polymerization will give low viscosity at high resin solids content, favoring rapid wetting and impregnation of the paper substrate, while the high proportion of methylol groups will give it fast cross-linking and curing capabilities.

A second, equally successful approach is to produce an MF resin of lower methylol group content and higher degree of condensation to which a small second addition of melamine (typically, 3 to 5% of total melamine) is effected toward the end of resin preparation. The shift to lower viscosity and higher solids content given by a second addition of melamine, shifting to lower values the average of the resin molecular mass distribution, yields a resin of rapid impregnation characteristics. Conversely, the higher degree of polymerization of the major part of the resin gives fast cross-linking and curing, due to the lower number of reaction steps needed to reach the gel point. Typical total M/F molar ratios used in this system are 1:1.5 to 1.7.

Figure 3.3 shows typical temperature and pH diagrams for the industrial manufacture of MF resins. The important control parameters to take care of during manufacture are the turbidity point (the point during resin preparation at which addition of a drop of MF reaction mixture to a test tube of cold water gives slight turbidity) and the water tolerance or hydrophobicity point, which marks the end of the reaction. The latter is a direct measure of the extent of condensation of the resin and indicates the percentage of water on mass of liquid reaction mixture that the MF resin can tolerate before precipitating out. It is typically set for resins of higher formaldehyde/melamine ratios and lower condensation levels at around 170 to 190%, while for resins of lower formaldehyde/melamine molar ratios and higher condensation level it is set at around 120%. As can be seen from Fig. 3.3, the pH is lowered to 9 to 9.5 once maximum reaction temperature is reached, to accelerate formation of the polymer. Once the turbidity point is reached, the pH is again increased to 9.7 to 10.0 to slow down and more finely control the endpoint, determined by reaching the desired value of the water tolerance point. Industrial MF resins are generally manufactured to a 53 to 55% resin solids content, with a final pH of 9.9 to 10.4 (but lower pH values are used for low-condensation resins).

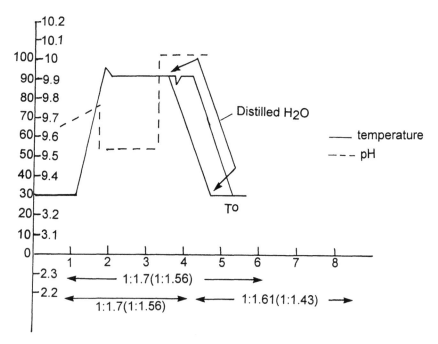

Figure 3.3 Typical manufacturing diagram for MF resins: y-axis, temperature in °C, and pH; x-axis, time in hours and variation of molar ratio.

If higher pH values are used, higher quantities of hardener need to be used, which is clearly uneconomical, in order to have acceptable rates of curing. For typical MF resins for low-pressure (particleboard) self-adhesive overlays, pressing times of between 30 and 60 s at 170 to 190°C press temperature are required according to the type of resin used. Pressing conditions for particleboard and plywood adhesives are identical to those used for UF resins.

B. Mixed Melamine Resins

As regards MUF, copolymers can be prepared which are generally used to cheapen the cost of MF resins but which also show some worsening of properties. Copolymerization was proven by means of model compounds and polycondensates [10]. MUF resins obtained by copolymerization during the resin preparation stage are superior in performance to MUF resins prepared by mixing preformed UF and MF resins, especially because processing of such mixtures is quite difficult [11]. The relative mass proportions of melamine to urea used in these MUF resins is generally in the range 50:50 to 40:60 of melamine/urea [12]. MPF resins, which in some respects show better properties of the corresponding MF and PF resins of origin, have also been prepared [13–15]. Analysis of the molecular structure of these resins in both their uncured and cured states showed that no co-condensates of phenol and melamine form and that two separate resins coexist. This is due to the difference in reactivity of the phenolic and melamine methylol groups as a function of pH. Also in their cured state, an interpenetrating network of the separate PF and MF resins, as a polymer blend, is formed, not a copolymer of the two [16–19].

III. PRINCIPLES OF PREPARATION OF MUF ADHESIVES

MUF adhesive resins have attracted attention as a very feasible alternative to the use of MF resins, because of the lower cost of the former. However, since the urea component of the resin confers reduced water resistance to the cured adhesive glue line, it has always been necessary to strike a balance among cost, performance, and durability in the case of MUF adhesives. First a product made from a MUF resin should contain as high a proportion of urea as possible, while limiting the consequent deterioration in the performance of the cured resin. In UF resins a series of step additions of urea has been found to yield resins of better performance [20,21]. A similar principle was applied to MUF resins to see if performance of the resin could be improved. In a mixed monomer resin several possibilities of step addition of the resin exist. The melamine or urea resin can be prepared first, followed by addition of one or several of the other components. The nature of the co-constituents formed during simultaneous reaction of urea and melamine with formaldehyde have been established [22] and the assignment of relevant ^{13}C NMR spectrum peaks has been reported [23,24,30]. As a step reaction should ensure systematic random condensation in blocks rather than simple random co-condensation of the monomer, resins prepared in the former manner are likely to have very different properties than those of the latter.

In a recent study [25] a few interesting conclusions were reached as regards to the best methods of manufacture and the conditions most important for preparation of a good MUF resin for particleboard. First, the method of manufacture must be considered. Formulations can be divided into those in which (1) the sequence of addition to the reactor is first melamine followed by first urea, followed by second urea (MUU); (2) those in which first and second urea addition are made before that of the melamine (UUM); and (3) those in which the initial reaction of urea and formaldehyde is followed by the addition of melamine and then of second urea (UMU). At low percentages of melamine in the total formulation, no differences between MUU and UMU were evident in the performance [25]. At medium percentages of melamine (50%), UMU formulations performed better than MUU formulations, in line with contemporary accepted wisdom for these resins. At higher melamine percentages (60%), MUU formulations perform better than UMU formulations, indicating that this is mainly a melamine resin on which urea is grafted.

It is also of interest to indicate at which molar ratio the resins' performance was best. At low melamine (M + U)/F molar ratios the resin gives the panel an optimum wet strength at 1:1.6. The dry strength of the board, on the other hand, keeps increasing with increasing proportions of formaldehyde. At medium percentages (50%) of melamine, while the same trend of improving dry strength with increasing formaldehyde proportions is evident, the wet strength improves to a plateau at 1:1.6 to 1.8 (M + U)/F molar ratio and no longer appreciably improves or decreases with increasing formaldehyde proportions.

IV. MUF RESIN PREPARATION

MUF resins of exceptional performance, capable of producing particleboard under standard production conditions capable of satisfying the requirements of both the DIN V100 test and the CTBA V313 test, can be obtained by a simple stepwise reaction procedure [32,33]. This involves starting with the preparation of a UF resin, and at a certain stage of the reaction adding melamine, followed later by a further addition of urea [15]. The diagram of manufacture of such a resin is shown in Fig. 3.4, and a more detailed procedure for its preparation is described in Section VII. With this stepwise procedure, MUF with M/U mass ratios between 45:55 and 50:50 can be prepared whose performance is truly excellent.

V. CHEMICAL AND PHYSICAL ANALYSIS OF MF AND MUF RESINS

The analysis of MF and MUF resins is difficult when unknown products, particularly fully cured articles, have to be tested for UF and MF resins. Widmer [26] offered a method for the identification of UF and MF resins in technical products. This involves preparing crystalline products of urea and melamine and identifying them under the microscope. Melamine (in the form of melamine crystals) and urea (in the form of long, crystalline needles of urea dixanthate) can be seen. This method allows one to distinguish between urea and melamine even in a cured adhesive joint.

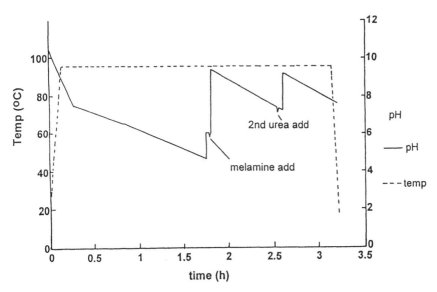

Figure 3.4 Typical manufacturing diagram for 40:60 to 50:50 M/U molar ratio MUF resins.

Quantitative determination of MF resins is also rather difficult. A method was developed by Widmer [26] for the quantitative determination of melamine in MF condensation products. In this method, the resins are destroyed under pressure by aminolysis, leaving the melamine intact. This is then converted to melamine picrate, which is easily crystallized and weighed. The Widmer method makes it possible to determine quantitatively the presence of urea and melamine in intermediate condensation products and in cured UF and MF resins (even when they have been mixed). Estimates are seldom in error by more than a few percent.

Hirt et al. [27] have published an effective and rapid method for the detection of melamine by ultraviolet spectrophotometry. This method can be used for products containing MF. It makes use of the strong absorption of the melamine ion at 235 nm. The resin is extracted from comminuted MF samples by hydrolyzing it to melamine by boiling it under reflux in 0.1 N hydrochloric acid. Stafford et al. [28] also provide a method for the identification of melamine in wet-strength paper.

Uncured MF resins analysis are carried out by gel permeation chromatography (GPC) and ^{13}C NMR. GPC is an inconvenient method for MF resins. Mostly dimethylformamide, dimethylsulfoxide, or salts solutions are used as solvents, with a differential refractometer as detector. Derivatives of the resin, such as those obtained by silation, are generally used to decrease molecular association by hydrogen bonding. ^{13}C NMR is a more convenient technique and the chemical shifts of the different structural groups in the resin can easily be identified [29,30]. This method is also quite convenient in comparing MF resins structures obtained by different manufacturing methodology.

A. ^{13}C NMR Spectra

In a recent article [25] use was made of ^{13}C NMR spectra of liquid MUF resins to try and forecast the quality of the resin in its hardened state. This procedure has been applied successfully to UF resins [20,21]. Assignment of ^{13}C NMR peaks for both UF and MF resins are well known [31] and have also been applied to predict the qualities of particleboard and laminated panels made from MF resin [25]. Figures 3.5 to 3.7 show MUF resins all made by the same procedure and contrast the spectra of resins, which were of widely diverging quality. The strong signal at $J = 40$ ppm is the solvent peak.

Figure 3.5 shows a resin of 60:40 melamine/urea content and (M + U)/F molar ratio 1:2.0 [25]. The high-field region of the spectrum between $J = 66$ and 77 ppm shows a dearth of peaks, which would indicate potential branching or cross-linking points. Only a single peak at $J = 65$ ppm (for an –NHCH$_2$OH

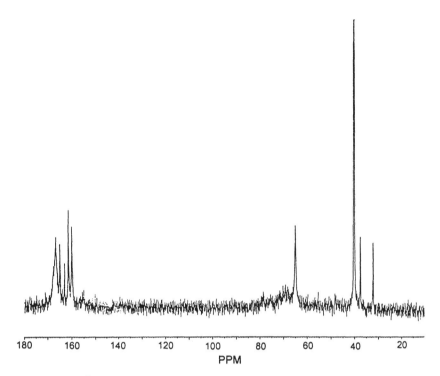

Figure 3.5 ^{13}C NMR spectrum of an "overcooked" resin of 60:40 mass M/U and (M + U)/F molar ratio of 1:2.

group) is of any importance. The peaks in the low-field region, $J = 160$ to 170 ppm, show that the hydroxymethylation has proceeded well—the secondary melamine peak at $J = 165$ ppm is almost as high as that of the unreacted melamine at $J = 166$, while the two peaks (at $J = 161$ and $J = 159.9$ ppm) for singly and doubly substituted urea are much stronger than that for unreacted urea at $J = 163$ ppm. This is an example of an "overcooked" resin, showing that curing is hindered when the condensation is allowed to proceed too far. A probable reason for this is the deterioration in the physical (a poor flow from high viscosity) and chemical (an overabundance of –NHCH$_2$OH groups with an accompanying deficit of bridging and branching groups) properties of the resin solution.

Figure 3.6 shows a much better quality resin of (M + U)/F molar ratio 1:1.8 and M/U mass ratio of 47:53 [25]. The ratio of the primary to secondary triazine peaks is 0.59 (compared to 1.10 of the previous resin), and for the

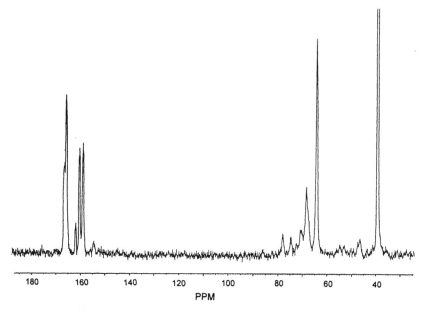

Figure 3.6 ^{13}C NMR spectrum of a MUF resin of 47:53 M/U mass ratio and (M + U)/F 1:1.8 molar ratio.

secondary and tertiary urea peaks to the primary urea peak is 0.16. A small doublet peak at J = 154.5 ppm shows the presence of co-condensed melamine and urea residues. In the high-field region, the peak for the hydroxymethyl group (–NHCH$_2$OH) is unusually strong at J = 64 ppm, but very prominent peaks at J = 68 ppm (methylene ether bridges) and at J = 70 (showing a hydroxymethyl group attached to a bridging group) testify to the good quality of the resin.

Another good-quality resin of M/U mass ratio 47:53 and (M + U)/F molar ratio 1:1.6 is shown in Fig. 3.7 [25]. This reaction is also well advanced, with the ratio for the triazine peaks being 0.46 and that for the urea peaks being 0.11, indicating high hydroxymethylation product yields, confirmed by the strong peak at J = 65 ppm for a –NHCH$_2$OH group. It is important that this peak should be prominent but should be of the same approximate height as the highest peak in the high-field region (J = 160 to 170 ppm). If it is higher, this usually occurs at the expense of the peaks denoting potential branching points and bridging units (J = 68 to 80 ppm) which would tend to signify a poor-quality resin.

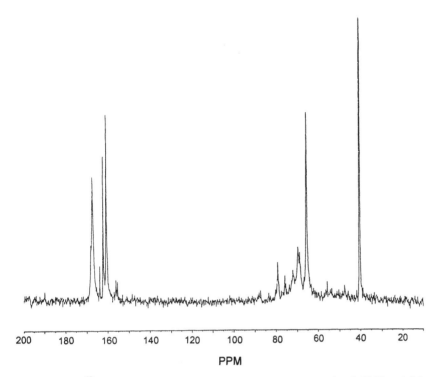

Figure 3.7 ^{13}C NMR spectrum of a MUF resin of M/U mass ratio of 47:53 and (M + U)/F molar ratio of 1:1.6.

Quantitative determination of MF- and MUF-bonded particleboard internal bond (IB) strength from peak ratios of the liquid resin ^{13}C NMR has been obtained [31] on lines similar to those described for UF resins. MF and MUF are somewhat more difficult cases in this respect, particularly the latter, as mixed melamine–urea peaks must also be considered in the relevant equations. A very simple relationship, which is more useful, can also be derived from the ^{13}C NMR spectrum of a liquid MUF resin to determine the unknown M/U ratio of a MUF resin [31]. In this regard a feature of the ^{13}C NMR spectrum of MUF resins is that the triazine ring carbon atoms and the urea carbonyl carbon differ in their response of peak intensity to concentration. The relationship which was then found to determine experimentally the M/U molar ratio was [31]

M/U ratio = 0.443 log(peak height ratio)^3 + 0.53(peak height ratio) + 1.44

with a standard deviation of 0.4. This is only an approximate equation, of course, but it can be of definite use.

The ^{13}C NMR spectrum of a resin solution can then be used to forecast the physical properties of the cured resin produced from it, and second, indicates that the reaction procedure can be as important as the urea content in determining the properties of the cured resin formed [24]. In particular, the procedure used, in which the reaction order was urea first addition, melamine, and urea second addition, produced significant improvement over the other reaction procedures in quality of particleboard [24,25]. Third, although an increase in the proportion of melamine out of the total melamine and urea content from 26% to 50% resulted in a definite improvement in the properties of the resin. The increase from 50% to 60% is not accompanied by any appreciable improvement in properties that would justify the increased cost of such a product [25,33]. The resins presented all have too much emission of formaldehyde. Such resins, however, can be used to good effect in conjunction with glue mix techniques to decrease formaldehyde emission, as discussed in Chapter 2 [34].

VI. GLUE MIXING

Glue mixing presents different requirements according to the final use of the MF resin. Hardeners are either acids or materials that will liberate acids on addition to the resin or on heating. In MF and MUF adhesives for bonding particleboard and plywood, the use of small percentages of ammonium salts, such as ammonium chloride or ammonium sulfate, is well established and is indeed identical to standard practice in UF resins. In MF adhesives for low- and high-pressure self-adhesive overlays and laminates, the situation is quite different. Ammonium salts cannot be used for the latter application for two main reasons: first, evolution of ammonia gas during drying and subsequent hot curing of the MF-impregnated paper would cause high porosity of the cured MF overlay; and second, the stability of ammonium salts, in particular of ammonium chloride, might cause the MF-impregnated paper to cure and deactivate at ambient temperature after a short time in storage, causing the resin to have lost its adhesive capability by the time it is needed in hot curing. Also, the elimination of ammonia during drying and curing would leave the cured, finished paper laminate essentially very acid, due to the residual acid of the hardener left in the system. This badly affects resistance to water attack of the cured MF surface, defeating the primary advantage for which such surfaces have justly become so popular. Thus a stable, self-neutralizing, non-gas-releasing hardener is needed for such an application. Several have been prepared. One of the most commonly used is the readily formed complex between morpholine and *p*-toluenesulfonic acid. Morpholine and *p*-toluenesulfonic acid readily react exothermically to form a complex of essentially neutral pH, which is stable up to well above 65°C.

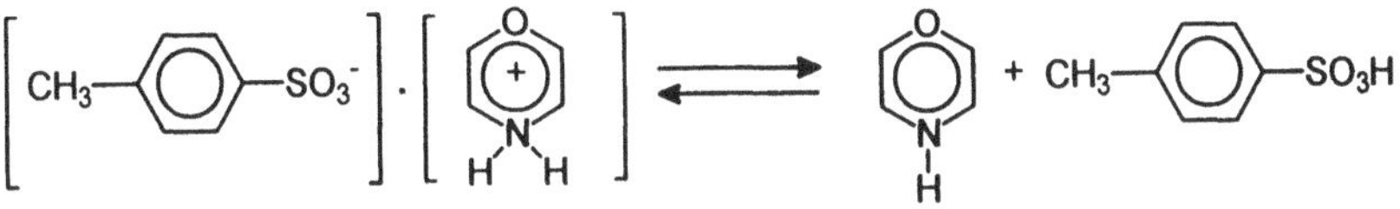

During heat curing of the MF paper overlay in the press, the complex decomposes, the MF resin is hardened by the acid that is liberated, morpholine is not vaporized and lost to the system, and on cooling the complex is re-formed, leaving the cured glue line essentially neutral.

In MF glue mixing for overlays and laminates, small amounts of release agents to facilitate release from the hot press of the cured bonded overlay are added. Small amounts of defoamers and wetting agents to further facilitate wetting and penetration of the resin in the paper are always added. A typical glue mix is shown in Table 3.1.

VII. FORMULAS

A. MUF Resins

For starting to experiment in MF resins, the two following formulations are suggested:

1. Formulation for Exterior Particleboard

In a reaction vessel charged at 25°C, to 44.4 parts by mass of water, are added 30% NaOH to pH 11.2 to 12.0, followed by 15.5 parts of 91% paraformaldehyde prills, 34.8 parts of melamine powder, 2.8 parts of

Table 3.1 Typical Paper Impregnation Glue Mix (Parts by Mass) for Self-Adhesive Low-Pressure MF Overlays

MF resin, 53% solids content	99.1
Release agent	0.08
Wetting agent	0.16
Hardener (morpholine/*p*-toluenesulfonic acid complex)	0.64
Defoamer	0.02

caprolactam, and 2.5 parts of N,N'–dimethylformamide while maintaining the temperature at 25°C, followed by heating in ±40 min to 92 to 95°C. When the temperature reaches 80°C, the pH is adjusted with 30% NaOH solution, if necessary, to pH 9.9. At 93°C, the mixture is cooled to 90°C and the temperature maintained there. The pH is adjusted to 9.55 to 9.65 with formic acid. The pH is held at this value while checking it, adjusting it, and recording it every ±10 min. The turbidity point is checked at 10-min intervals until the turbidity point is reached. At this time the pH is brought up to 9.95 to 10.05. The pH checked, adjusted, and recorded every 10 min. Water is distilled under vacuum to a solid of ±53 to 55%. Water tolerance is checked at 10-min intervals until it is 170 to 180%. Then full vacuum is applied and the resin is cooled to 30 to 35°C.

2. Formulation for Low-Pressure MF-Paper-Impregnated Overlays

The same procedure is followed as that for formulation 1 above, but at the end of the water vacuum distillation 1.7 to 1.9 parts by mass of second melamine are added and the reaction mixture is heated up to 95°C again, and this temperature is maintained for 5 to 6 min. Then the mixture is cooled rapidly.

B. MUF Resins

1. Urea–Melamine–Second Urea (UMU) (Figs. 3.4 and 3.7)

To 113 parts of Formurea (a formaldehyde concentrate stabilized by urea of mass content 57% formaldehyde and 23% urea) are added 13 parts urea and 30 parts water. The pH is set at 10 to 10.4 and the temperature brought to 92 to 93°C under mechanical stirring. The pH is then lowered to 7.8 and the reaction continued at the same temperature, allowing the pH to fall by itself over a period of 1 h 30 min to 1 h 35 min to a pH of 5.2 (the pH must never fall below 5). To bring the pH to 9.5 or higher, 22% NaOH was added, then 41 parts of melamine premixed with 19 parts of water. One part of dimethylformamide and 2 parts of diethylene glycol are then added to the reaction mixture, maintaining a temperature of 93°C. The water tolerance is checked every 10 min while the pH is allowed to fall by itself. When the water tolerance reached is 180 to 200% (this was reached after 35 to 40 min, and the pH reached is 7.2), 6.5 parts by mass of second urea is added and the pH is again brought up to 9.5. The reaction is continued until the water tolerance reached is lower than 150% (the pH has reached 7.7 at this stage).

The pH is then corrected to 9.5 again and the reaction mixture cooled and stored. Resins produced using this procedure have a solids content of 58 to 65%; a density of 1.260 to 1.280 at 20°C, a viscosity of 70 to 150 cP, free formaldehyde of 0.32, and gel times with 3% NH_4Cl of 51 to 57 s at 100°C. The preparation diagram of this resin is shown in Figure 3.4.

REFERENCES

1. A. Pizzi, Chapter 2 in *Wood Adhesives: Chemistry and Technology*, Vol. 1 (A. Pizzi, ed.), Marcel Dekker, New York, 1983.
2. R. Koehler, *Kunst. Tech.*, *11*: 1 (1941); *Kolloid Z.*, *103*: 138 (1943).
3. R. Frey, *Helv. Chim. Acta*, *18*: 491 (1935).
4. K. Sato and T. Naito, *Oikym. J.*, *(Jpn.)*, *5*(2): 144 (1973).
5. M. Akano and Y. Ogata, *J. Am. Chem. Soc.*, *74*: 5728 (1952).
6. K. Sato and S. Ouchi, *Polym. J.* *(Jpn.)*, *10*(1): 1 (1978).
7. A. Takahashi, *Chem. High Polym.* *(Jpn.)*, 7: 115 (1950); *Chem. Abstr.*, *46*: 438 (1952); *Chem. High Polym.* *(Jpn.)*, 9: 15 (1952); *Chem. Abstr.*, *48:* 1730 (1954).
8. *Synthetic Resins Adhesives for Wood*, British Standard B5 1204, 1965.
9. J. M. Dinwoodie, Chapter 1 in *Wood Adhesives: Chemistry and Technology*, Vol. 1 (A Pizzi, ed.), Marcel Dekker, New York, 1983.
10. D. Braun and H.-J. Ritzert, *Angew. Makromol. Chem.*, *156*: 1 (1988); *135*: 193 (1985).
11. D. Braun and H.-J. Ritzert, *Kunststoffe*, *77*: 1264 (1987).
12. W. Clad and C. Schmidt-Hellerau, *Proceedings of the 11th International Particleboard Symposium*, Washington State University, Pullman, Wash., 1977, pp. 33–61.
13. A. Bachmann and T. Bertz, *Aminoplaste*, VEB Verlag fur Grundstoffindustrie, Leipzig, 1967, p. 81.
14. A. Knop and W. Scheib, *Chemistry and Application of Phenolic Resins*, Springer-Verlag, Berlin, 1979, p. 134.
15. K. Bruncken, in *Kunststoffhandbuch* (R. Vieweg and E. Becker, eds.), Carl Hanser, Munich, 1968, Vol. 10, p. 352.
16. D. Braun and W. Krausse, *Angew. Makromol. Chem.*, *108*: 141 (1982).
17. D. Braun and W. Krausse, *Angew. Makromol. Chem.*, *118*: 165 (1983).
18. D. Braun and H.-J. Ritzert, *Angew. Makromol. Chem.*, *125*: 9 (1984).
19. D. Braun and H.-J. Ritzert, *Angew. Makromol. Chem.*, *125*: 27 (1984).
20. E. E. Ferg, A. Pizzi, and D. Levendis, *J. Appl. Polym. Sci.*, *50*: 907 (1993).
21. E. E. Ferg, A. Pizzi, D. Levendis, *Holzforsch. Holzverwert.*, in press (1993).
22. R. E. Kirk and D. F. Otmer, *Encyclopedia of Chemical Technology*, Interscience, New York, 1947.
23. J. R. Ebdon, B. J. Hunt, T. S. O'Rourke, and J. Parkin, *Br. Polym. J.*, *20*(4): 327 (1988).

24. T. A. Mercer, M.Sc. thesis, University of the Witwatersrand, Johannesburg, South Africa, 1993.

25. T. A. Mercer and A. Pizzi, *Holzforsch. Holzverwert.*, in press (1993).

26. G. Widmer, *Paint Oil Chem. Rev.*, *112*: 18, 26, 28, 30, 32–34 (1949); *Kunststoffe*, *46*(8): 359 (1956).

27. R. C. Hirt, F. T. King, and R. G. Schmitt, *Anal. Chem.*, *26*(8): 1273 (1954).

28. R. W. Stafford, *Paper Trade J.*, *120*: 51 (1945).

29. H. Schidlebauer and J. Anderer, *Angew. Markromol. Chem.*, *79*: 157 (1979).

30. B. Tomita and H. Ono, *J. Polym. Sci. Chem. Ed.*, *17*: 3205 (1979).

31. T. A. Mercer, Ph.D. thesis, University of the Witwatersrand, Johannesburg, South Africa, 1994.

32. Spanplatten: Flach-pressplatten für das Bauwesen, DIN 68763 (1982).

33. *Standard for Particleboard*, Centre Technique du Bois et de l'Ameublement (CTBA), Paris.

34. A. Pizzi, L. Lipschitz, and J. Valenzuela, *Holzforschung*, *48* (3):226 (1994).

4

Phenolic Resin Wood Adhesives

I. INTRODUCTION

A. Chemistry

Phenolic resins are the polycondensation products of the reaction of phenol with formaldehyde. Phenolic resins are the first true synthetic polymers to be developed commercially. Polyfunctional phenols may react with formaldehyde in both the ortho and para positions of the hydroxyl group. This means that the condensation products exist as numerous positional isomerides for any chain length. The characteristic that renders these resins invaluable as adhesives is their ability to deliver, at a relatively low cost, water, weather, and high-temperature resistance to the cured glue line of a joint bonded with phenolic adhesives.

Phenols condense initially with formaldehyde in the presence of either acid or alkali to form a methylol phenol or phenolic alcohol, and then dimethylol phenol. The initial attack may be at the 2-, 4-, or 6-position. The second stage of the reaction involves the reaction of the methylol groups with other available phenol or methylol phenol, leading first to the formation of linear polymers [1] and then to the formation of hard-cured, highly branched structures.

Novolak resins are obtained with acid catalysis, with a deficiency of formaldehyde. A novolak resin has no reactive methylol groups in its molecules, and therefore is incapable of condensing with other novolak molecules on heating without hardening agents. To complete resinification, additional formaldehyde is added to cross-link the novolak resin. Phenolic rings are considerably less active as nucleophilic centers at an acid pH, due to hydroxyl and ring protonation. However, the aldehyde is activated by protonation, which compensates for this reduction in potential reactivity. The protonated aldehyde is a more effective electrophile. The substitution reaction proceeds slowly and condensation follows as a result of further protonation and the creation of a benzylcarbonium ion, which acts as a nucleophile.

Resols are obtained as a result of alkaline catalysis and an excess of formaldehyde. A resol molecule contains reactive methylol groups. Heating causes the reactive resol molecules to condense to form large molecules without the addition of a hardener. The function of phenols as nucleophiles is strengthened by the ionization of the phenol, without affecting the activity of the aldehyde.

The difference between acid- and base-catalyzed processes is (1) in the rate of aldehyde attack on the phenol, (2) in the subsequent condensation of the phenolic alcohols, and (3) to some extent, in the nature of the condensation reaction. With acid catalysis, phenolic alcohol formation is relatively slow. Therefore, this is the step that determines the rate of the total reaction. The condensation of phenolic alcohols and phenols forming compounds of the dihydroxydiphenylmethane type is, instead, rapid. The latter are therefore predominant intermediates in novolak resins.

In the condensation of phenols and formaldehyde using alkaline catalysts, the initial substitution reaction (i.e., the formaldehyde attack on the phenol) is faster than the subsequent condensation reaction. Consequently, phenolic alcohols are initially the predominant intermediate compounds. These phenolic alcohols, which contain reactive methylol groups, condense either with other methylol groups to form ether links, or more commonly, with reactive positions in the phenolic ring (ortho or para to the hydroxyl group) to form methylene bridges. In both cases water is eliminated.

Mildly condensed liquid resols, which are the more important of the two types of phenolic resins in the formulation of wood adhesives, have an average of less than two phenolic nuclei in the molecule. The solid resols average three to four phenolic nuclei but with a wider distribution of molecular size. Small amounts of simple phenol, phenolic alcohols, formaldehyde, and water are also present in resols. Heating or acidification of these resins causes cross-linking through uncondensed phenolic alcohol groups, and possibly also through reaction of formaldehyde liberated by breakdown of the ether links. As with novolaks, the methylolphenols formed condense with more phenols to form methylene-bridged polyphenols. The latter, however, quickly react in an alkaline system with more formaldehyde to produce methylol derivatives of the polyphenols. In addition to this method of growth in molecular size, methylol groups may interreact with one another, liberating water and forming dimethylene–ether links CH_2–O–CH_2. This is particularly evident if the formaldehyde/phenol ratio is high. The average molecular weight of the resins obtained by acid condensation of phenol and formaldehyde decreases hyperbolically from over 1000 to 200, with increases in the molar ratio of phenol to formaldehyde from 1.25:1 to 10:1. Other important aspects of phenolic resin chemistry, such as acid and chlorine catalysis and metallic ion acceleration at mildly acid and neutral pH values, have been discussed in depth elsewhere [1].

B. General Principles of Manufacture

A typical phenolic resin is made in batches in a jacketed, stainless steel reactor equipped with an anchor-type or turbine-blade agitator, a reflux condenser, vacuum equipment, and heating and cooling facilities. Molten phenol, formalin (containing 37 to 42% formaldehyde or paraformaldehyde), water, and methanol are charged into the reactor in molar proportions between 1:1.1 and 1:2, and mechanical stirring is begun.

To make a resol-type resin (such as that used in wood adhesive manufacture), an alkaline catalyst, such as sodium hydroxide, is added to the batch. It is then heated to 80 to 100°C. Reaction temperatures are kept under 95 to 100°C by applying vacuum or by cooling water in the reactor jacket. Reaction times vary between 1 and 8 h, according to the pH, the phenol/formaldehyde ratio, the presence or absence of reaction retarders (such as alcohols), and the temperature of the reaction.

Since a resol can gel in the reactor, dehydration temperatures are kept well below 100°C, by applying vacuum. Tests have to be done to determine first the degree of advancement of the resin, and second, when the batch should be discharged. Examples of methods of such tests are measurement of the gel time of a resin on a 150°C hot plate or at 100°C in a water bath. Another method is by measurement of the turbidity point by precipitating the resin in water or solutions of a certain concentration.

Resins that are water soluble and of low molecular weight are finished at as low a temperature as possible, usually around 40 to 60°C. It is important that the liquid, water-soluble resols retain their ability to mix with water easily since when they are used as wood adhesives they often require the addition of water to counterbalance the effect of the fillers added. Resols based on phenol are considered to be stable for 3 to 9 months. Properties of a typical resin are a viscosity of 100 to 200 cP at 20°C, a solids content of 55 to 60%, a water mixibility of a minimum of 2500%, and a pH of 7 to 13, according to the application for which the resin is destined.

C. PF Wood Binders

Phenolic resins are used as binders for exterior-grade plywood and particle-board, which need the superior water resistance provided by these resins. In the manufacture of plywood, the phenolic resin adhesive is usually applied to the wood veneers by roller or extrusion coating. The coated veneer is then cross-grained, stacked, and cured in a multi-daylight press for 5 to 10 min at 120 to 130°C at 11 to 16 kg cm^{-2}. In the manufacture of particleboard, the phenolic resin adhesives are sprayed onto the wood chips, or are sprayed and spread by continuous blenders. The glued wood chips are formed in a mat and then pressed for 20 to 12 s/mm^{-1}, according to thickness, press temperature, and moisture content, at 180 to 230°C and 25 to 35 kg cm^{-2}.

The only type of phenolic resins commercially used for this application are resol resins. These are hardened by heating after the addition of small amounts of wax emulsion and insecticide solution in the case of particleboard, and of vegetable or mineral fillers and tackifiers in the case of plywood. The pH of these resins varies between 10 and 13.5 and is generally between 12 and 12.5. In the case of particleboard, typical pressing times of PF resins are 12 to 17 s per millimeter thickness of the board at 200°C press temperature. These pressing times are much slower than for amnioplastic resins and constitute the most obvious drawback for the use of PF wood adhesives.

II. PF–LIGNOCELLULOSIC INTERFACE AT MOLECULAR LEVEL

The adhesion of PF [2–4] and UF oligomers [5,6] to cellulose as well as the adhesion of other resins, such as acrylics, to a generalized substrate [7] has been modeled via molecular mechanics by means of the sum of secondary interaction forces between the various oligomers constituting a resin and the substrate. The correspondence between theoretical conclusions and experi-mental evidence which has been obtained confirms that the molecular mechanics approach used in these cases appears to be well suited to describing the phenomenon of adhesion between a substrate and the macromolecules constituting a synthetic resin.

The results of the energy interaction of the five methylol phenols (hydroxybenzyl alcohols) and the three PF dimers (dihydroxydiphenyl-methanes) obtained in the initial condensation stages of phenol with formal-dehyde (Figs. 4.1 and 4.2) indicate that there are significant differences in the values of the minimum total energy.

The first noticeable difference (Table 4.1, Fig 4.2) from what has been observed for UF resins [5,6] is the different role that the introduction of methylol groups appears to have in the case of PF condensates. In the latter, introduction of the methylol groups appears initially to decrease the energy of adhesion. Thus from Fig. 4.2 (schemes 1 and 2) it is apparent that the values for the average of the minimum energies of interaction for the sum of the different sites (scheme 1), the minimum energies of interaction (scheme 2), and even the minimum energies divided by the number of atoms in each molecule (in parentheses in scheme 2, Fig 4.2) decrease when passing from phenol to the two monomethylol phenols and then to the two dimethylol phenols. Thus in the initial PF condensates up to the dimethylol phenols, the energy of adhesion decreases with the introduction of an increasing number of methylol groups. This is the reverse trend to what is seen in UF resins, where, instead, the introduction of an increasing number of methylol groups appears to increase the energy of adhesion [5,6]. Trimethylol phenol does not maintain this trend, except for presenting a lower value than phenol for the minimum of energy divided by the number of atoms in each molecule (values in parentheses, scheme 2, Fig. 4.2). These differences between the energies of adhesion of methylolated species between PF and UF oligomers are then noticed in the PF phenol–monomethylolated–dimethylolated species sequence, while the dimethylol to trimethylol and monomethylol to trimethylol energy trends are comparable to those observed in UF condensates [5,6]. These differences may be ascribed to the increased steric hindrance that the introduction of an increasing number of methylol groups in a phenol molecule may cause. Phenol is a much bulkier molecule than urea, and hence two opposite trends develop when introducing methylol groups: (1) the increased attraction caused by introducing a very polar group such as the methylol group is counteracted by (2) the increased steric hindrance due to the bulkier molecule in which the methylol group is introduced, denying the possibility of finding a conformation of lower energetic interaction with the substrate. Thus the fitting and conformational minimization of the methylolated phenolic species to the cellulose substrate is much more difficult when steric hindrance is enhanced by introduction of a methylol group. A clear indication that steric

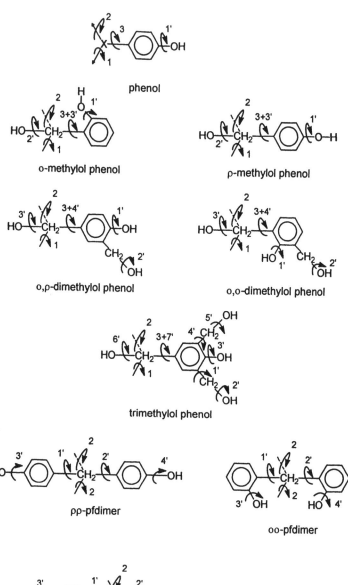

phenol

o-methylol phenol

p-methylol phenol

o,p-dimethylol phenol

o,o-dimethylol phenol

trimethylol phenol

ρρ-pfdimer

oo-pfdimer

oρ-pfdimer

Figure 4.1 PF species studied on crystalline cellulose.

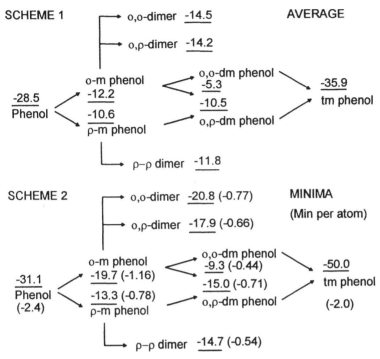

Figure 4.2 Schematic flowcharts of the PF species formed in the initial stages of the reaction of phenol and formaldehyde, with their total energies of adhesion (scheme 1), adhesion energy minima (scheme 2), and average energies of adhesion per atom of each molecule (scheme 2, in parentheses).

hindrance is the cause of these differences is shown by the total value of the hydrogen bonds. The contribution of hydrogen bonds to the total attractive energy is much lower in the methylolated species that in the PF dimers. As regards trimethylol phenol, the number of methylol groups is so high that the attractive effect again begins to predominate. This would not, however, improve adhesion in the initial condensation stages of a PF resin because at any stage in a PF condensation the relative proportion of trimethylol phenol in relation to the monomethylolated and dimethylolated species is indeed very small. Equally, a trend different from that observed in UF resins is observed when passing from the mono- and dimethylolated species to the non-methylolated dimers. In the PF oligomer series, attraction increases, whereas in the UF oligomer series, it decreases. The computational results obtained are shown in Fig. 4.2.

Table 4.1 Comparison of Computational Energy Results with Experimental
Paper Chromatography R_f Values for PF Hydroxybenzyl Alcohols (Methylol
Phenols) and Dihydroxydiphenylmethanes (PF Dimers), Showing Qualitative
Correspondence of Their Relative Positions

| Species | Interaction energy (kcal mol^{-1}) | | R_f [8–10] | R_f |
	Average	Minimum		
Trihydroxybenzyl alcohol	−35.9	−50.0	0.34	—
o,o-Dihydroxydiphenylmethane	−14.5	−20.8	—	0.55
o,p-Dihydroxydiphenylmethane	−14.2	−17.9	—	0.55
2-Hydroxybenzyl alcohol	−12.2	−19.7	0.84	0.85
p,p-Dihydroxydiphenylmethane	−11.8	−14.7	—	0.65
2,4-Dihydroxybenzyl alcohol	−10.5	−15.0	0.61	0.60
4-Hydroxybenzyl alcohol	−10.6	−13.3	0.79	0.80
2,6-Dihyroxybenzyl alcohol	−5.3	−9.3	0.67	0.68
2-Hydroxybenzyl alcohol/Na$^+$ complex	−9.9	−12.3	0.84	0.85

[a]Experimental results from Refs. 8 and 9.

The computational energy values for the five hydroxybenzyl alcohols and
three dihydroxydiphenylmethanes are correlated with the R_f values obtained
by paper chromatography (Table 4.1). It can be seen that the sequence in
which the eight compounds present themselves on the paper chromatogram
correlates well qualitatively with the sequence of both their average energy
and minimum energy values. There are, however, two exceptions in the paper
chromatography results reported by Freeman [8,10]: 2-hydroxybenzyl alcohol
and 2,6-dihydroxybenzyl alcohol.

Of these two compounds, the one whose relative minimum and average
energies appear to be completely out of line with the experimental R_f values
is 2-hydroxybenzyl alcohol. Such a compound is in the presence of sodium
hydroxide, which always dissolves it in water, and in the preparation of PF
condensation. Caesar and Sachanen [10,11] have postulated the formation of
a chelate group involving the alcohol, the phenolic hydroxy groups, and the
sodium in PF alkali-catalyzed condensations according to the scheme.
Evidence for the formation of such ring complexes has been reported
[12,50,52]. Thus the computation was repeated for the Na ring complex of
the hydroxybenzyl alcohol. The results are reported in Table 4.1. In the case
of such complexes being formed, the relative R_f value of the 2-hydroxybenzyl
alcohol correlates well with the result calculated for its Na complex.

The only compound left out of line is the 2,6-dihydroxybenzyl alcohol. Controversy has occurred as to the relative R_f value of this compound on paper chromatography, and some authors [13] have placed it with second highest R_f value, while others have tended to confirm the results of Freeman [14]. Suggestions that such a compound is in a form stabilized by a double intramolecular hydrogen bonding [8–10], and the possibility that Na complexes might also form in this case, render it particularly difficult to understand what molecular species is really present, and thus it is not possible to guess which molecular configuration has to be used computationally for this case.

Of even greater interest is the more in-depth comparison of the relative energies of interaction, hence of the adhesion, between each of the PF dimers and an elementary crystallite of cellulose. There is an 18.5% difference in energy between the para–para and ortho–para dimers, and 2.2% between ortho–para and ortho–ortho [3], (para–para isomers, -10.203 kcal mol^{-1}; ortho–para, -12.085 kcal mol^{-1}; ortho–ortho, -12.329 kcal mol^{-1}) when the configuration of minimum energy of interaction is minimized and averaged over 24 sites of the cellulose crystallite [3]. Experimental paper chromatography results reported R_f differences for the three dimers of 18.2% and 2%, respectively; these values compare very favorably with the values of the minimum conformation averages attained by computational molecular mechanics methods. It confirms the accuracy of molecular mechanics methods in describing adhesion to a lignocellulosic substrate when a molecular model has been chosen. These results were of interest not only in the adhesive–substrate interaction field but were the first case demonstrating the feasibility of modeling the separation of nonenantiomeric isomers on an achiral cellulose substrate [3]. Modeling of the resolution of enontiomers from racemic mixtures (on crystalline and amorphous cellulose) had been achieved successfully shortly before for totally different systems than for adhesives [15–17]. There is no doubt that application of the new systems of molecular dynamics computations [18] (rather than molecular mechanics) will advance this field of application further.

Of more immediate applied interest are the consequences of such molecular mechanics studies. Among these is the interesting concept [2] that the adhesion, expressed as energy of interaction, of a PF oligomer with cellulose is greater than the average attraction of cellulose for water molecules [19,20]. This also implies that PF oligomers are likely to displace water to adhere to a cellulose surface. Caution should be exercized in understanding that this is valid only for certain attractive sites and is not an easily generalized conclusion. It is, nonetheless, an interesting deduction because it is an important factor in wood bonding: the adhesion of the synthetic resin to wood should be considerably better than the adhesion of water molecules to the same substrate. It is particularly important, first, for "grip" by the adhesive of the substrate's surface, and second, in the cured adhesive state, in partly determining the level of resistance to water attack of the interfacial bond between adhesive and adherend. The greater average attraction of cellulose sites for PF oligomers rather than water is a definite contributory factor to PF-bonded materials being classed as weather and water resistant; it is not only due to the imperviousness to water of the cured PF resin itself but also to the now theoretically identified imperviousness to water of the adhesive–adherend interfacial bond.

The most interesting finding of a molecular mechanics approach is the difference in energy of adhesion between ortho–ortho and ortho–para oligo-mers in relation to para–para oligomers [2]. In general, the distribution of methylene linkages in commercial PF resins indicates a higher percentage of ortho–ortho and para–para linkages over ortho–ortho linkages. The higher calculated adhesion of ortho–ortho and ortho–para indicates that a shift in manufacturing procedure of the resin to minimize the proportion of para–para linkages would result in a resin capable of better adhesion to the substrate [2]. This can be achieved in several ways, such as by introducing ortho-ori-entating additives to produce "high-ortho" PF resins [1,12,50,51]. The implication is that a decrease in adhesive content without loss of performance in a product such as particleboard is possible just by a simple shift in the resin manufacturing procedure.

III. COVALENT BONDING VERSUS CATALYTIC ACTIVATION OF PF CURING BY WOOD SURFACES

A complete understanding of the phenomenon of adhesion of a phenolic, or any other resin, to a lignocellulosic substrate such as wood is fundamental to the evaluation of bond strength and bond strength development in wood products. Calculations based on molecular mechanics [2–7] and dynamics [18] can only explain rather statically the relationship between synthetic resins and wood constituents at the glue-line interface. The influence that the substrate exercises on the kinetic and curing behavior of the resin itself is not immediately obvious with the former approaches. It has been definitely established repeatedly, for instance, that the energy of activation of the reaction of polycondensation of PF resins, and also of UF and MF resins, is influenced markedly by the presence of wood [21–26]. In the presence of wood the energy of activation of the polycondensation reaction of these resins is lowered considerably. This implies that resin polymerization and cross-linking proceed at a much faster rate when the resin is in molecular contact with one or more of the wood constituents. This phenomenon has, since its initial discovery, been ascribed to the formation of covalent bonds between synthetic resin and the main constituents of the wood matrix. It has led to the widespread concept that excellent wood bonding can be achieved only when a high proportion of covalent bonds is formed between synthetic resin and the main polymeric constituents of the wood matrix. Although there is no doubt that covalent bonds can be formed linking both PF resols and UF and MF resins with lignin and wood carbohydrates, it is important to consider the conditions under which they are formed. Prolonged heating and very high temperatures are conducive to covalent bonds being formed, a fact clearly demonstrated by certain applications of these resins to natural textiles [27,28]. The controversy that needs to be addressed, then, is not if covalent bonds are or are not formed, but instead, (1) if they are formed, and (2) to what extent, if any, they are formed, under standard application and reaction conditions for resins used as thermosetting wood adhesives. The maximum temperature in a particleboard core, for example, only reaches a maximum of 100 to 110°C at the pressing times and temperatures characteristic of such a wood bonding process [1]. Pressing times vary between 7 and 15 s per millimeter of board thickness. Neither temperatures nor times are then extreme. Instead, the conditions under which the formation of resin–substrate covalent bonding has been noted are characteristic of differential scanning colorimetry investigations, with scanning increments of 10 to 20°C/min, up to maximum temperatures of 250 to 300°C [21–24,26]. Even considering the main curing exotherm of a PF resin as occurring at approximately 150°C (which it does), this will be reached in not less than 5 to 10 min after heating has started. At the pressing times used, a particleboard core remains at a maximum of 100 to 105°C for only 20 to 60 s, and reaches this maximum in 80 to 120 s.

A variety of investigations have been conducted to demonstrate the concept of resin–substrate covalent bonding. The majority of these have been by differential scanning colorimetry (DSC) and thermogravimetric analysis (TGA) [21–24,26]. Infrared (IR) spectrophotometry and ultraviolet (UV) spectrophotometry have also been used to demonstrate the formation of covalent bonding. The time and temperature limitations affecting DSC and TGA investigations have been outlined briefly above. The IR and UV investigations used even more extreme conditions: 14 h at 190°C for the IR [29,30] and 2 h at 120°C for the UV investigation [25]. As a consequence, the formation of covalent bonds is demonstrated, but the extreme conditions used also imply clearly that the situation is very different at the much lower temperatures and much shorter times characteristic of wood adhesive applications. A controversial but acute review [31] of the subject concluded that the evidence presented clearly indicates that no covalent bonds are formed under thermosetting wood adhesive applications. This might also be too extreme. The molecular mechanics findings outlined earlier give a strong theoretical grounding to the theory that strong wood adhesive bonding can be explained equally well and completely by the sum of a multitude of resin–substrate secondary force interactions [2–7] without any need for interfacial resin–substrate covalent bonds. In this the molecular mechanics conclusions are in line with advanced theories of adhesion in adhesives fields other than for wood [32]. The undeniable experimental fact remains, however, that a lignocellulosic substrate facilitates PF and other adhesive polymerization by lowering the energy of activation of the polycondensation.

It was left to a very recent series of investigations [33,34] to shed more light on the covalent versus secondary bonding controversy in the wood adhesives field. In these investigations not only were the energy of activation of a standard PF resin on wood and on each of its constituents determined by DSC, but also the reaction of simple model compounds of the PF resin such as o- and p-hydroxybenzyl alcohols. The DSC findings were backed by a concomitant TGA investigation, by model compounds kinetics, and by IR data [33,37].

To better understand the nature of the findings, a few statements of a general nature are needed. Differential scanning colorimetry is a technique recognized to give results that are, at best, difficult to interpret: this is a well-known argument to specialists in the DSC field [35]. It is for this reason that such a technique is often used exclusively for quick (just indicative) comparative means. First, the energies of activation obtained automatically from the DSC equipment for each scan are completely inconsequential, and hence they cannot be relied upon even for purely comparative purposes. Second, shifts in maximum temperature and in intensity of an exotherm cannot be used, as it is believed instead, to deduce the higher or lower reactivity of a certain resin in the presence of a certain substrate: they might mean something and equally often do not mean anything. A serious DSC investigation involves calculation of the reaction's energy of activation by the Kissinger equation [36]. This equation relates the ln of the heating rate (β) divided by the square of the maximum of temperature of the exotherm (T_{max}) as a function of the inverse of the maximum of temperature reached by the exotherm being examined.

$$\ln \frac{\beta}{T_{max}^2} = -\frac{Ea}{RT_{max}} + \ln \frac{AR}{E} + C$$

From the Kissinger equation, after a series of scans at various heating rates, if the data are reliable, a linear relationship is observed between $\ln \beta/T_{max}^2$ and $1/T_{max}$. From this and this alone, the correct energy of activation of the reaction corresponding to the exotherm being studied can be calculated. It does not even end at this. A series of exotherms in a DSC scan can be real or fictitious: transformations in the substrate, such as oxidations and degenerative modifications, can be the cause of some of the apparent exotherms. Checking for substrate reactions alone is therefore a necessity. Checking the temperatures at which each exotherm starts and ends is also necessary. Checking the temperature range of the endotherms is also of importance: after all, in polycondensations such as for PF, UF, and MF resins the polycondensation proceeds by liberation of water. Thus if an exothermic peak is not followed by an endothermic one, the exotherm is not likely to be real, or if real, it is not likely to belong to the polycondensation of the resin with itself or of the resin with the substrate. It is clear, then, that DSC data alone can be quite unreliable if the identical experiment is not backed by some other technique. One of the most apt is TGA and its first derivative: this allows the identical experiment, under experimental conditions identical to those used in a DSC scan, to be conducted. The TGA analysis gives a step trace of the decrease in mass of the sample under heating. In a PF resin polycondensation reaction, for each condensation step, water is released and evaporated with a concomitant decrease in mass of the sample. Coupling a DSC trace with a TGA trace (or its first derivative) rapidly and efficiently gives an idea of which DSC exotherms are due to any type of polycondensation, which are not, and which are fictitious.

The recent investigation mentioned addressed the above for both PF resins and two model compounds [33,34]. The model compounds' DSC scans yielded a series of several exotherms, a gratifying result if one considers that a monomeric monohydroxybenzyl alcohol must self-condense to dimers first and then to higher oligomers in a series of steps, each being exothermic. The PF resin instead gave a series of only three exotherms. In Figs. 4.3 and 4.4

the DSC scans of the PF resin alone and of *p*-hydroxybenzyl alcohol alone are shown with their corresponding TGAs and derivatives. These are shown to give the reader a feeling for the type of data obtainable. The final summarized energies of activation of the study are reported in Table 4.2. The three exotherms of the PF resin show correspondence with those of the model compound: the first PF exotherm with the first exotherm of the model compound, the second PF exotherm with the sum of the second and third exotherms or the third exotherm only of the model compound, and the third PF exotherm a wider example of the fourth column exotherm of the model compound. The study concluded that all the constituents of the wood substrate, at one stage or another of the reaction, contribute to decrease the energy of activation of the entire polycondensation or of important steps in it. The data, as presented in Table 4.2, cannot show directly if the decrease in energy of activation is due to covalent coreaction between resin and substrate, or if it is caused by a catalytic activation of the resin self-condensation induced by the wood constituents.

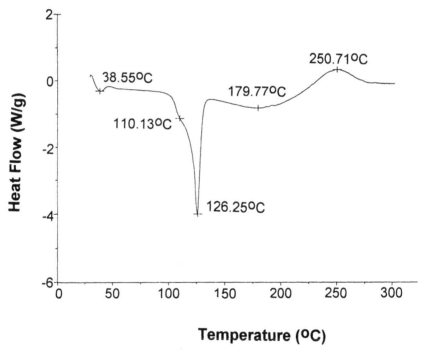

Figure 4.3   DSC scan of *p*-hydroxybenzyl alcohol.

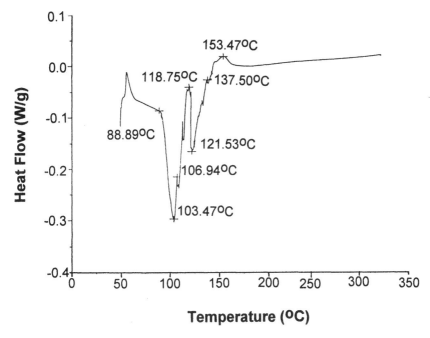

Figure 4.4   DSC scan of PF resin at 5 s/mm.

Some interesting aspects are immediately noticeable, however, and important conclusions can be drawn. Any real exotherm that is not present in
the self-condensation of the model compound alone is probably due to
reactions between resin and substrate. This is so because any reaction between
model and substrate will be a different condensation case than self-condensation of the model; it will give a totally differently positioned exotherm
specific to, and characteristic of, that particular reaction. Table 4.2 makes it
clear that covalent bonds between resin and lignin do form, and at a variety
of stages in the polycondensation. That some covalent bonds between model
compound (and model resin) and cellulose form at the higher temperatures is
also evident. Also evident is the fact that the energy of activation of the
characteristic self-condensation exotherms of the model compound on some
substrates, particularly cellulose at the lower temperatures, also decreases
substantially. This cannot be ascribed to condensation exotherms exclusively
specific to reaction with the substrate of the model resin. This is an interesting
deduction, clearly implying that the substrate catalytically activates the
self-condensation of the model and model resin. Thus, because catalytic

Table 4.2 Activation Energies (kcal mol^{-1}) of PF Polycondensation and Curing Obtained by DSC and Kissinger Equation, Corresponding to TGA Mass Losses, for PF Resin and Hydroxybenzyl Alcohol Models

	Hydroxybenzyl alcohol models						
p-Hydroxybenzyl alcohol alone	25.24	24.58	31.18	30.14	—	26.83	—
p-Hydroxybenzyl alcohol + lignin	—	72.06	25.89	—	20.09	12.62	—
p-Hydroxybenzyl alcohol + cellulose*	—	—	12.67	—	21.56	23.39	36.71
p-Hydroxybenzyl alcohol + wood flour	29.75	60.56	30.48	—	—	21.31	15.04
Average peak temperature (°C)	98.4	110.7	139.7	149.9	194	247.6	267.2
Peak temperature range (°C)	93.1–103	107.6–115.7	127–155.9	145.8–152	180.6–210.3	231.7–261	256.5–280
Main exotherm peak				Main*			

	PF resin models		
PF resin alone	55.61	23.85	29.15
PF resin + lignin	—	15.82	21.37
PF resin + cellulose[a]	6.91	16.73	20.84
PF resin + cellulose[b]	20.34	20.21	26.29
PF resin + wood flour	14.53	12.04	8.44
Average peak temperature (°C)	90.1	135.4	161.7
Peak temperature range (°C)	65–104	102.8–159	147.9–179.9
Main exothermic peak		Main	

[a]Cotton wool.
[b]Filter paper.

activation of the resin–substrate covalent bonds being formed is characteristic, self-polycondensation and curing of a PF resin on a lignocellulosic substrate also occur. At the lower temperatures of greater interest for wood adhesive applications, both lignin and particularly, crystalline cellulose, appear to activate catalytically PF polycondensation and curing [33,34].

The question that remains, then, is: To what extent is wood bonding independent of resin–substrate covalent bonds being formed, or of resin–substrate secondary forces being present, the latter also being the driving cause of the resin-accelerated self-condensation [2–6,33,34]? The extent and intensities of the exotherm, or better, the area enclosed under the DSC exotherm peaks, indicate this. Thus the main exotherm of the PF resin (Table 4.2 and Fig 4.3) due to PF self-condensation based on activation of the polycondensation constitutes well over 90% of the total area included under all the exotherms and well over 95% of the area included under the lower-temperature exotherms. Comparative values are obtained for saligenin and *p*-hydroxybenzyl alcohol models, although important differences between the *o*- and *p*-isomeride models occur.

Unusual results also occur; for example, the first exotherm of the PF resin proper in Table 4.2 indicates no peak in the PF + lignin case, indicative that in the PF resin used, any reaction with lignin in that particular temperature range occurs to an extent too small to yield a visible, detectable exotherm. In this context, lignin appears to retard the reaction—only in appearance, though, this apparent result probably being due to its poorer wettability by the resin, hence due to a diffusion problem. Instead, cellulose appears to again activate the reaction, and the value of the results for wood flour confirm this in relation to the proportion of cellulose and crystalline cellulose contained in the wood sample [33,34]. The study used softwood wood flour (*Pinus radiate*), the guaiacyl character of the lignin of which is typically between 85 and 95%. In hardwoods, and in latewood, in which syringyl moieties are generally in higher proportion, the occurrence of resin–substrate covalent bonding should become more remote.

In conclusion, this DSC study indicated that the proportion of exotherms due to resin–substrate covalent bonding appears to be low in relation to the surface activation exotherms [33,34]. Unequivocal supporting evidence for this being the case has also been obtained from several other sources and approaches:

1. TGA traces indicate that mass loss is small in the endotherms corresponding to a preceding covalent bonding exotherm, while it is huge for the endotherms following self-condensation of the resin [33,34] (Figs. 4.3 and 4.4).

2. Model compound kinetics [33,34] of the reactions of phenol, guaiacol, veratrol, and glucose at 95°C, in the pH range 2 to 10.5, indicate clearly that the rate constants obtained always favor especially the reaction of formaldehyde (and PF methylols) with phenol. This is followed fairly closely by the reaction of formaldehyde with guaiacol, which at 95°C is ±15% slower than the former. The reaction of veratrol (1,2-dimethoxybenzene) and glucose with formaldehyde presents rate constants 230 and 550 times smaller, respectively, than the PF reaction [33,34] (Table 4.3). Covalent bonding between resin and substrate can then occur readily on guaiacyl lignin nuclei presenting a free phenolic hydroxy group, assuming that there are no problems of diffusion [33,34]. The resin–substrate covalent bonding reaction is definitely too slow to occur on lignin guaiacyl nuclei in which a phenolic hydroxyl is not present (i.e., veratrol). The typical number of free phenolic hydroxyls in guaiacyl lignins is reported to be low. Hence the proportion of covalent bonds that can form between resin and substrate has to be small under any reaction conditions, and very small indeed under the standard conditions prevalent in thermosetting wood adhesive application. The reaction leading to covalent bonding between resin and wood carbohydrates is much less favorable, considering the rate constant of the glucose–formaldehyde reaction; a clear confirmation of the preponderance of the catalytic activation effect of some carbohydrates on PF resin self-condensation [33,34].

3. Reflectance Fourier transform (FT)-IR investigations on PF resins and model systems also confirmed the DSC findings, with the formation of only a very small proportion of covalent bonds being detected in reaction times of up to 2 h [33,34].

4. Other studies [37,38] have also attempted to determine the possible extent of covalent bonding of a thermosetting adhesive to cellulose under

Table 4.3 Rate Constants of the Initial Condensation Reaction of Formaldehyde with Simple Model Components of a Phenolic Resin (Phenol) and of Wood Constituents

	k (L mol^{-1} s^{-1})
Phenol–formaldehyde	3.3×10^{-2}
Guaiacol–formaldehyde	2.9×10^{-2}
Veratrol–formaldehyde	1.4×10^{-4}
Glucose–formaldehyde	5.8×10^{-5}

conditions closely resembling those during UF particleboard hot pressing (160°C, 3 to 6 min pressing, acid catalysis). Cotton was treated with an aqueous solution of N,N'-dimethylolurea at NH_4Cl concentrations of 0 to 2%, subsequently dried at 115°C and heated at 150°C for 10 min. The amount of nonwashable material permanently bound to the cotton varied from 2 to 6% of the original dimethylolurea, depending on NH_4Cl concentration [37,38]. Again a very clear indication that the very significant decrease in energy of activation induced in a resin by cellulose is due primarily to catalytic activation of the polycondensation rather than by the very small proportion of resin–substrate covalent bonding. In the specific case of a PF resin's reaction on cellulose, the study of Allan and Neogi [39] is of particular significance. After 12 h at 80°C under nitrogen, they concluded that any alkaline reaction of hydroxybenzyl alcohols (methylol-phenols) with α-cellulose occurs only very infrequently: the proportions given can be summarized as 1.6 covalent bonds in 100 units for lignin and 2 covalent bonds in 1000 units for cellulose [39].

5. Molecular mechanics results at the resin–substrate interface explain well the physicochemical phenomenon of bonding just by secondary force interactions between resin and substrate [2–7].

The reasons for the existence of catalytic activation induced by the substrate for the adhesive resin self-condensation must be considered. Molecular mechanics has indicated that the sum of secondary forces binding resin oligomers to substrate is quite considerable [2–4]. Figure 4.2 gives values for a variety of PF oligomers, and Fig. 2.5 for UF oligomers. As secondary forces involve electronic, charge, and dipolar interactions, these strong forces of attraction to the substrate cause variations in the strength of bonds and intensity of reactive sites within the PF oligomer considered. This is well known in heterogeneous catalysis for a variety of chemical systems [40] other than the one at hand. Bonds cleavage and formation within a molecule or between molecules are greatly facilitated by chemisorption onto a catalyst surface [40], as illustrated in Fig. 4.5.

In the case of PF resin/cellulose, and possibly of UF and MF resins too,

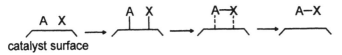

catalyst surface

Figure 4.5 Heterogeneous catalysis: schematic representation of a catalyst surface on facilitating bond formation between two molecules, A and X.

the strong secondary interactions weaken some bonds in the phenolic oligomer molecule. Taking as an example the case of a monomethylolot PF dimer such as

the bonds that are most likely to be weakened are the C–O bond of the methylol group, inducing a much stronger positive charge on the carbon atom, possibly even a carbocation. Conversely, further weakening of the phenolic O–H bond will lead to a stronger negative charge delocalized on the ortho and para reactive sites of the aromatic ring. The self-condensation reaction, a bimolecular reaction, is then much more rapid as a consequence of the substrate-induced activation of the only two types of reactive sites of the phenolic oligomers. A secondary force attraction of 20 kcal mol^{-1} between synthetic oligomer and substrate, for example, if concentrated primarily on a single bond such as the C–O of the methylol group, would weaken it, hence activate the reaction up to 25%. In theory, the reaction of two PF molecules activated both to such an extent would yield a condensation up to a purely theoretical maximum ±600 times faster than the same reaction without the substrate. This means, for example, that a PF condensation rate constant of 3.3×10^{-2} L mol^{-1} s^{-1} (Table 4.3) at 95°C would, with surface activation, have a rate constant of 1.98×10 under identical reaction conditions. Calculation of the decrease in energy of activation, under an equivalent collisional factor (which is doubtful), would give a decrease in energy of activation from 23.85 kcal mol^{-1} (Table 4.2) down to a value of 19.1 kcal mol^{-1}: approximately the same order of magnitude observed in Table 4.3 for cellulose surface activation of the main curing exotherm. Variations in collisional factors expected in such a different system, and diffusion considerations are likely to account for the remaining difference (Table 4.2) [40].

As a final argument on this subject, it must be noted that for particleboard and other bonded particulate panels, some distinction must be made between the board core and board surfaces. The case of the former has already been presented. In the case of board surfaces, in direct contact with the press hot platens, the temperature reached is much higher. A higher proportion of covalent bonding across the resin–substrate interface could then be expected.

This is still likely to be very small, however, as the UF experiments of Poblete and Roffael [38] and Allan and Neogi [39] clearly indicate.

An interesting concept [33,34] which can be advanced is that some covalent bonding can form, this being the initial step to catalytic activation of the resin self-condensation, as shown in Fig. 4.5. This is difficult to ascertain, although such a concept would also explain the low proportion of resin–substrate covalent bonds found at any time, these being only part of an intermediate, transient state. It is likely that the covalent bonds formed on lignin are not of this type, but that they are stable, whereas those found on cellulose might be part of such a transient mechanism. With the present data, it is not possible to say.

IV. PF ADHESIVES ^{13}C NMR RELATION TO RESINS CURED STRENGTH

As in the case of UF and MF resins, relationships between various ^{13}C NMR peak intensities ratios of a PF adhesive resin in its liquid state and the strength of the adhesive in its cured, hardened state can be developed [41,42]. This evolves into equations correlating ^{13}C NMR peak ratios of the resin directly with the internal bond strength of the particleboard bonded with it [41,42]. The advantage of such a method is of course comparative, as various PF resins can be compared in terms of their performance characteristics without having to pass through the board manufacturing stage. Identification of the ^{13}C NMR peak shift from characteristic chemical groups of a PF resin is a well-researched field, with significant contributions by several researchers [43–45].

It is this knowledge that has allowed the correlation of ^{13}C NMR peak ratios of a PF resin with dry and wet internal bond strength of particleboards bonded with it, under standard conditions of preparation [41,42]. As for aminoplastic resins, in PF resins all chemical groups are closely interrelated, and ratios of the intensities of peaks characteristic of chemical groups which are known or suspected to contribute to the cured strength appear to indicate excellent correlation with experimental reality for PF adhesives [41,42].

With this approach a general equation can be written relating the extent of cured resin strength and the internal bond of a board bonded with it to certain ^{13}C NMR peak ratios [41,42].

$$\text{IB strength (MPa)} = a\,\frac{A}{A+B+C} + b\,\frac{Mo}{A+B+C} + c\,\frac{ME}{A+B+C}$$

where A = sum of peak intensities of ortho and para sites still free to
           react on phenolic nuclei (110 to 122 ppm)
       B = meta sites on phenolic nuclei, sum of peak intensities (125
           to 137 ppm)
       C = sum of peak intensities of already reacted ortho and
           para sites (125 to 137 ppm)
       Mo = methylol groups, sum of peak intensities (59 to 66 ppm)
       Me = methylene bridges, sum of peak intensities (30 to 45 ppm)

The equation above gave an excellent correlation between IB strength of a
PF-bonded particleboard and the liquid PF resin ^{13}C NMR spectrum [41,42]
(Fig. 4.6). More interesting was the finding that in PF resins, correlation
between IB strength and ^{13}C NMR is even better than for UF resins. A
conceptually less correct but simpler equation than the above but also giving
good correlation was also developed, this being [41,42]

$$IB\ strength\ (MPa) = d\,\frac{A}{A+B+C} + e\,\frac{Me}{Mo}$$

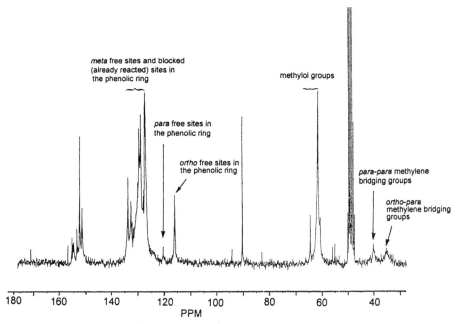

Figure 4.6   PF resin ^{13}C NMR spectrum, with relevant peaks for calculation of IB
strength highlighted.

a, b, c, d, and e in the two equations being coefficients characteristic of the type of manufacturing procedure used for the PF resin. The equations above were found to be predictive in the phenol/formaldehyde molar ratio range from 1:1.1 to 1:2.2 [41,42]. As an example of the types of values obtained for the coefficients, a series of resins such as resin 1 (lower-condensation PF resin) and a stepwise reagent addition resin (higher-condensation PF resin) (resin 1) presented in Section XI at the end of this chapter gave the following values:

One-step low-condensation resins

$$\text{IB dry (MPa)} = -0.470 \frac{A}{A+B+C} + 1.431 \frac{Mo}{A+B+C} + 3.734 \frac{Me}{A+B+C}$$

standard deviation + 0.04

$$\text{IB wet (V100)} = -0.743 \frac{A}{A+B+C} + 0.428 \frac{Mo}{A+B+C} + 1.915 \frac{Me}{A+B+C}$$

standard deviation = 0.07 MPa

Multistep high-condensation resins

$$\text{IB dry (MPa)} = -0.134 \frac{A}{A+B+C} + 1.874 \frac{Mo}{A+B+C} + 3.734 \frac{Me}{A+B+C}$$

standard deviation + 0.02 Mpa

$$\text{IB wet (V100)} = -0.134 \frac{A}{A+B+C} + 0.967 \frac{Mo}{A+B+C} + 0.059 \frac{Me}{A+B+C}$$

standard deviation = 0.006 MPa

In the case of these PF resin equations, one immediately noticeable difference from the UF resin equations is the absence of the "pre-cross-linked" term, the term given by tridimensionally branched units in the liquid resin, which is, instead, of considerable weight in UF resins [46,47]. This again confirms that in their liquid, solution form, PF resins for wood bonding are linear and that no noticeable amount of prebranching is present. Furthermore, the coefficient of the term A/(A + B + C) is always negative, this indicating that the higher the amount of still free ortho and para sites, the lower the degree of condensation of the resin, and potentially, the lower the density of cross-linking at parity of still available methylol groups.

As for UF adhesives, an equation relating ^{13}C NMR peak ratios of the PF resin with formaldehyde emission of the particleboard bonded with it has been derived [41,42,46]. This equation is of less importance than that for UF resins, as formaldehyde emission from PF-bonded wood products is in general not problematic, being well below statutory limits [48]. It might gain more importance in the future as statutory limits are lowered.

Form, emission

$$= a \frac{Mo}{A+B+C} + b \frac{free\ formaldehyde}{A+B+C} + c \frac{methylene\ ethers}{A+B+C}$$

For simple low-condensation, one-step resins, the coefficients a, b, and c were found to be −39.4, +38.2, and +105.0, respectively; for multistep higher-condensation resins, the coefficients a, b, and c were found to be 417.9, +2.44, and −43.9, respectively. The standard deviation on the results obtained was found to be 0.410 and 0.887 mg of formaldehyde per 100 g of board [41,42].

V. BEHAVIOR OF PF RESINS UNDER VERY ALKALINE CONDITIONS

The effect of pH on the rate of curing and polymerization of phenolic resins is well known [1,49] (Fig. 4.7). It is also widely accepted that at alkaline pH the rate of curing of phenolic resins accelerates [1,49]. This occurs as the function of phenolic nuclei as nucleophiles is strengthened by ionization of the phenol to form phenate ions [1,49]. The literature, however, is lacking in information regarding exact data on the slope of the gel time (hence reactivity) curves as a function of pH, over a pH of 9 [1,10,49]. This is an unusual gap if it is considered that a great proportion of the total world output of PF resins is used as thermosetting wood adhesives, an application in which only PF resins of very high alkalinity, generally at pH values between 10 and 13, are used to impart faster reactivity and shorter pressing time to the adhesive.

The curves of gel time versus pH in Fig. 4.8 illustrate the unexpected behavior of Na^+-catalyzed PF resins [50]. Over a pH range of approximately 9 to 10, the rate of curing of the PF resin slows down markedly, instead of accelerating as always supposed. It continues to slow down markedly with increasing pH. In 1948, Caesar and Sachanen [11] postulated that the phenolic hydroxy group must be greatly involved in the reaction of the methylol group to form methylene linkages in alkaline phenol–formaldehyde resins. They based their hypothesis on the comparison between the impossibility of reacting tiophene and formaldehyde in alkaline environment with the ready reaction by phenol and formaldehyde under the same reaction conditions. Their hypothesis was only a postulate, without any proof, and was disregarded and criticized at the time. It was, however, considered interesting and unusual

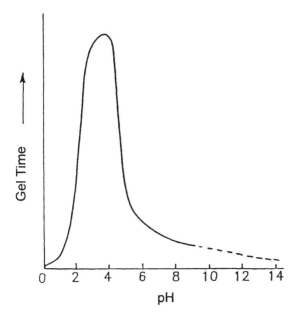

Figure 4.7   Gel time versus pH curve of PF resin. Dashed line indicates extrapolated assumption.

enough to be reported by Megson [10] and advanced again as a possible explanation for NaOH catalysis by Fraser et al. [12,51]. Caesar and Sachanen postulated the formation of an intramolecular chelate group holding the Na^+ between the phenolic and the o-methylol hydroxyls of a PF resin. They carried their idea too far by again postulating exchange of the Na^+ ion from the phenolic hydroxy group to the methylol group to form Na^+–OCH_2^- ions, a very unlikely possibility indeed. Justifiably, the criticism of their ideas stemmed primarily from the latter assumption. It is indeed very unlikely that an alcoholic group can behave as an acid when a much more acid group such as the phenolic hydroxyl is present. For instance, the acceleration or re-tardation of PF resins in an acid, neutral, and even mildly alkaline environment by bivalent or trivalent ions such as Zn^{2+}, Ba^{2+}, Cr^{3+}, and others was first reported by Fraser et al. [12,51] and ascribed to ring mechanisms similar to those proposed for Na^+. Much later, their hypothesis was demonstrated by the isolation and characterization of several of these ring complexes [52] and by the determination of their kinetics of acceleration and retardation of PF resins [53]. The mechanism proposed for the progressive retardation of PF

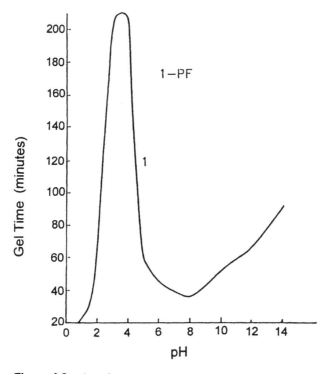

Figure 4.8 Actual experimental gel time versus pH curves of a PF resin alone, showing the upward trend of the gel-time curve.

resins that appear to be consistent with what is observed and all the rest of the evidence is then

The mechanism would be likely to occur both in an already formed PF resin, as shown above, in which the methylol group attached to the aromatic nuclei is already present, and in the reaction of phenol and formaldehyde in the presence of NaOH as originally postulated by Caesar and Sachanen [11].

The presence of this mechanism and of the ring complex would explain with ease the progressive retardation of PF resins curing at increasingly higher pH values [50,54]. Freeman [55] and Freeman and Lewis [56] determined the relative reactivities of individual nuclear positions in phenol and methylol phenols under alkaline reaction conditions (see above). From these it is clear that the rate of curing of a PF resin is mostly independent of the o-hydroxybenzyl alcohol reactive sites present in the PF resin, p-hydroxybenzyl alcohol reaction sites being too slow to be the determining factor in the rate of cure of the resin. Thus any structural rearrangements, even small, affecting the electronic structure of o-hydroxybenzyl alcohols are likely to have a noticeable effect on the rate of curing of the resin. Once the Na^+ ring complex is formed, two factors affect the reactivity of the o-hydroxybenzyl alcohol moieties in the resin: first, the methylol group is blocked, or its mobility decreased and its capacity to react to methylene linkages and cross-linkages is affected; and second, the strong inclusion of Na^+ in the complex decreases the electron density on the phenolic aromatic ring, markedly reducing the carbanion negative charge, which is the main force driving the PF condensation reaction under alkaline conditions. Thus, as the pH increases, a higher proportion of methylol groups are blocked (or better, they would show a decrease in their collisional factor) and a higher proportion of carbanions present a less negative charge, slowing the reaction down, as observed in the gel-time curves presented. The Na^+ complex ring can be stable only at higher pH values and is likely to be in equilibrium with free Na^+ ions present in the solution. At lower pH values the amount of Na^+ in solution is not enough to maintain the equilibrium sufficiently shifted toward the ring complex to ensure its stability [50,54].

In practice it has proved impossible to isolate the Na^+ complex or to confirm its existence definitely, as it would be stable only in solutions of high Na^+ concentration. Direct and indirect proof for its existence and against its existence has, however, been obtained. Paper chromatography of PF oligomers, in particular of all the hydroxybenzyl alcohols, correlates well with molecular mechanics computational results of their interaction with the surface of crystalline cellulose [3]. The R_f sequence of all the hydroxybenzyl alcohols on a paper chromatogram correlates well with the sequence of their values of both minimum and average energy of interaction with the cellulose surface obtained by molecular mechanics computation [3]. There is, however, one exception to this trend in all the paper chromatography results reported for hydroxybenzyl alcohols by Freeman [55,56] and other authors [57–60]. The exception is saligenin (o-hydroxybenzyl alcohol), the surface interaction energy of which appears to be completely out of line with its experimental R_f value on paper chromatography. It has the highest experimental R_f of all the hydroxybenzyl alcohols indicating that its energy of interaction with the substrate surface should be one of the lowest, while the molecular mechanics computational results indicate that it should be the second slowest-moving compound [4,50]. However, saligenin is always in the presence of some sodium hydroxide which is necessary to dissolve in water. Molecular mechanics computations indicate that the Na^+ ring complex of saligenin would be quite stable. It also indicates that its secondary forces interactions with the cellulose surface are exactly in line with the experimental R_f value observed for saligenin paper chromatography [4,50]. Again, a clear indication that saligenin in very alkaline pH might be present in the form of the proposed Na^+ complex. Intramolecular H-bonding of saligenin without Na^+ increases its computational R_f but only by about half of what is needed, and thus its participation could be discarded.

There is further, but not determining, proof that only the o-hydroxybenzyl alcohol moieties of PF resins are the ones involved in the retardation effect. Firstly, at pH 13.5–14 the gel time curves [50,54] (Fig. 4.8) indicate a gel time which is only, approximately, 45% of that obtained at the pH of minimum reactivity (pH 3.5–4) inferring that only some types of reactive groups and sites might be involved in the retarding effect (but this would also be valid for other mechanisms) [50,54]. Secondly, gel time curves of aged PF resins, hence resins of higher degree of polymerization and lower methylol group content, show a shift in the onset of the retarding effect from pH 8.5–9 to pH 10–10.5 and a less marked upward trend of the gel time curve. Aging decreases the number and proportion of the more reactive carbanions sites and methylol groups, those of the o-hydroxybenzyl alcohol moieties: the retarding effect will then appear later and be slightly less intense, and this has been observed.

Direct and negative proof, apparently denying the existence of this mechanism, is provided by UV spectrometry. If the Na^+ ring complex exists, possibly three species should be observed by UV at different pH values for o-hydroxybenzyl alcohol: the phenolic hydroxyl, the phenate ion, and the Na^+ ring complex. Only two are observed instead, exactly as for p-hydroxybenzyl alcohol and, even more damning, exactly as for *o*-cresol (*o*-methylphenol, in which the ring complex cannot exist). It could be argued that UV might not be able to distinguish between the alleged Na^+ ring complex and the noncyclical Na^+ phenate ion, but this evidence is damning nonetheless.

The second piece of indirect evidence, also apparently denying the existence of the Na^+ ring complex mechanism at such high pH values, is the behavior of PF resins prepared and catalyzed exclusively with KOH and with NaOH totally absent. The hypothesis of Bender [61] that KOH-catalyzed resins are faster due to the atomic radius of K^+ being greater than Na^+ and thus forming unstable K^+ ring complexes, or no complexes at all, was checked.

KOH-prepared and KOH-catalyzed PF resins indicate clearly that the differences between KOH- and NaOH-catalyzed resins do not warrant Bender's assumptions regarding the existence of a Na^+ ring complex. The existence of the Na^+ ring complex remains elusive, and the data presented appear to indicate that the ring complex is not formed because the only types of data appearing to indicate its existence are debatable. The R_f value of saligenin in paper chromatography could also be explained by intermolecular as well as intramolecular hydrogen bonding and by applying perhaps a more sophisticated molecular mechanics algorithm. Conversely, the difference in the onset of the ester accelerating effect in aged resins can be explained by the appearance of consistent amounts of quinone methides.

Any possible explanation for the strong cure-retarding effect shown by NaOH- and KOH-catalyzed PF resins at higher pH values is likely to lie in the formation of quinone methides [50,54]. Megson [10] states that quinone methides yield dibenzyl ether cross-links with ease and methylene bridges with great difficulty. Thus, in the ionic environment present in a 50% solution of a PF resin, as the proportion of quinone methides increases, the cross-linking condensation is likely to slow down. Quinone methides form with ease from ionic species; hence

where no negative charge is present, hence no carbanion-driven reaction occurs, equally where there is no positive charge to drive the condensation reaction. As the reaction is driven schematically by

the progressive rearrangement to quinone methides and their increasing proportion will progressively slow down the condensation step of the reaction, as is observed.

It is finally of interest to note if such an effect occurs only with phenol or also applies to reaction of other phenols with formaldehyde. The much more reactive resorcinol was used for this purpose, as ^{13}C NMR has shown that it reacts with esters. The effect is also present with the faster-reacting resorcinol, whose upward-trending gel-time curve, however, occurs at much higher pH values and is of much lower intensity. A similar trend is noticeable with higher-condensation PF resins, in which the cure retardation onset appears at higher pH values [50,54].

Particleboard results in which the same PF resin was used under standard laboratory/industrial application conditions but at different pH values show that the results are not consistent with the retardation effect noticed by the gel-time tests [54], although others do [62,63]. A good explanation for this unexpected behavior is the effect of the inherent acidity of timber. Pine timber is known to have a pH of around 4. The action of such acidity will be to lower the pH of the resin downward. The indications from the results obtained [54] is that the pH of 10% PF resin at pH 13 on timber falls to pH 10 or below: it is then near the pH of minimum reactivity. PF resins of lower pH will stabilize on timber even at lower pH, and thus their reactivity will fall in the slower part of the curve and give worse (but still acceptable) internal bond results.

In conclusion, the slowing down of gel times at high alkalinity does not affect the performance of the resin once this is on pine wood [54], due to the wood's inherent acidity, although on very acid woods this can happen [62,63].

It is of importance in applications where the amount of wood in relation to the resin is low or where wood is not used at all.

VI. ALKALINE CURING ACCELERATION METHODS FOR PF WOOD ADHESIVES

A. α-Set and β-Set [50,64]

The so-called α-set (in liquid phase) and β-set (in gas phase) acceleration of the cure of very alkaline PF resins for foundry core binders was pioneered in the early 1970s [65–67]. In this application the addition of considerable amounts of esters, such as propylene carbonate, methyl formate, and triacetin (glycerol triacetate), was found to reduce resin curing time to extremely short periods. This process is now used extensively worldwide for foundry core PF binders. This mechanism of PF curing acceleration can also be used for wood adhesive applications, such as for particleboard, in which the relevance of an adhesive is dependent on how fast a pressing time can be achieved for the panel.

The curves of gel times versus pH in Fig. 4.8 illustrate the slowing down of PF resin reaction with increasing pH. In Fig. 4.9 the increasing addition of an ester increasingly accelerates the reaction in the pH range under consideration. During reaction, hydrolysis of the ester occurs (gel times were carried out at 94°C), as noted by following the reaction by IR spectroscopy [50,64]. The effects of different esters and different amounts of esters on the curing acceleration of the same PF resin are shown in Fig. 4.9. The esters used in the study [50,64] were propylene carbonate, methyl formate, triacetin (glycerol triacetate), phenyl acetate, ethyl butyrate, and methyl salycilate. The strength and dissociation of the acid involved in the ester appeared to determine the rate of curing acceleration of the resin.

In more detail, the curves in Fig. 4.9b, in which the dependence of gel time on ester amount has been expressed as percentages of moles of acid generated by dissociation of the ester, indicate that at a constant amount of acid there is no difference between triacetin and phenyl acetate, both esters of acetic acid. Furthermore, the relative positions of the curves in Fig. 4.9b are on the order of the relative pK_a values of the acids in the esters. The only exception is propylene carbonate, in which the correlation between amount of ester and gel time is not linear, and the curve of which can be divided into two well-defined reactions: the first, up to approximately $2^1/2\%$ ester on PF resin solids, showing extremely fast curing acceleration, and a second, from

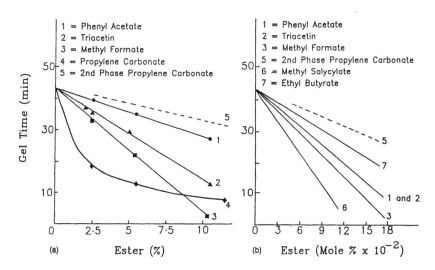

Figure 4.9   (a) Dependence of gel time of a PF resin at pH 11.26 from the percentage of ester added (percentage ester, by mass, on PF resin solids), and the type of ester added; (b) dependence of gel time of a PF resin, at pH 11.26 from the molar percentage of ester added on mass of PF resin solids, illustrating dependence on quantity of ester, pK_a of the type of acid in the ester, and effect of diprotic acid (for propylene carbonate).

approximately 5% ester, responding to the relative pK_a of the first hydrogen of carbonic acid (Fig. 4.9b).

Other interesting features noticeable in Fig. 4.10 are the very evident "bumps" in the pH range 8.5 to 11 observed in the curves of the ester-accelerated PF resins. In the PF curves without ester in Fig. 4.10, the same bumps, but less marked due to the upward trend of the curves, are noticeable in the same pH range. In the pH range 8 to 9.5 the methylol groups of a phenolic resin react readily to form metastable methylene ether cross-links [68]. It is the cleavage of methylene ether cross-linkages outside this pH range, with rearrangement to methylene linkages and liberation of formaldehyde [1,10, 68], which is likely to account best for the presence of these bumps in the curves.

Of all the ester accelerators, only propylene carbonate appeared to deviate from the established trend. Even taking into consideration both protons of carbonic acid, the fast gel times observed up to 2½ to 3% (Fig. 4.9b) could at best be reported in Fig. 4.9b near but not quite up to the slope of formic

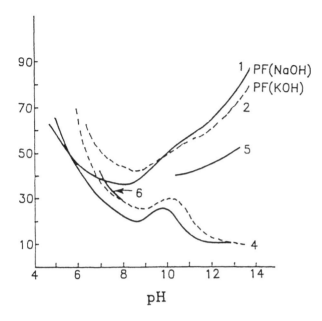

Figure 4.10 Cure retarding at high pH and ester acceleration effect of NaOH- and KOH-prepared PF resins (ester = propylene carbonate). 1, PF (NaOH); 2, PF (KOH); 3, PF (NaOH) + propylene carbonate (6.7%); 4, PF (KOH) + propylene carbonate (6.7%); 5, 4-month aged PF (NaOH); 6, PF (NaOH) + triacetin (6.7%). Note "bumps" caused by methylene ethers formation and decomposition at pH 8 to 11. Note curve 5, effect of 4-month aging of PF resins of curve 1 on the extent and starting pH value of the retardation effect. Compare the start of acceleration for curves 4 and 6, showing the difference between propylene carbonate and triacetin, and the starting point of acceleration at pH values of 5.5 and 7.1.

acid [50,64]. The mechanism of acceleration was found [50,64] to be based on an increase in the number of cross-linking sites in a phenolic resin. This is based on a fragment from the ester dissociation introducing a different, and additional, cross-linking mechanism other than that based on methylene formation [50,64].

Potentially, any ester should have this capability in very alkaline solutions; thus it is well known [1,10] that

Reactions such as (2) are known. For instance, in group transfer polymerization a very strong base is capable of doing just this [69].

In the very concentrated NaOH environment prevalent at very high pH values in a PF resin with a very reactive metastable compound such as propylenecarbonate, a similar carbanion attack may be possible [50,64]. From the relative reactivities of individual nuclear positions in phenol and methylol phenols under alkaline reaction conditions determined by Freeman [55] and Lewis [56], it is clear that phenolic carbanions [70] formed by o-methylol moieties in the PF resin will be those reacting first. The reaction of propylene carbonate with phenol, hydroxybenzyl alcohols, and PF resins then appears to be nothing but a variant of the Kolbe–Schmitt [71] synthesis (in which salycilic acid is obtained from phenol and carbon dioxide). The existence under very alkaline conditions of both OH$^-$ and phenolic carbanions will lead to competitive reactions determining the relative proportions of products obtained. A hypothesis regarding the mechanism of reaction has been advanced (Fig. 4.11) [50,64].

As for the Kolbe–Schmitt reaction it is the ortho position of the phenol, thus the ortho carbanion, which will react preferentially, although the para carbanion will also react. The same reaction as that for phenol alone would also occur with a linear PF resin in which extensive attack at the meta positions has been noted [50,64]. On this cross-linking mechanism the usual methylol-to-methylene cross-linking of PF resins is superimposed. Thus the second cross-linking mechanism, caused by propylene carbonate, is then also the probable case of ester acceleration of PF resins. Acceleration is observed (1) due to the presence of the equivalent of another trifunctional reagent (compare to difunctionality of formaldehyde for instance), (2) due to a

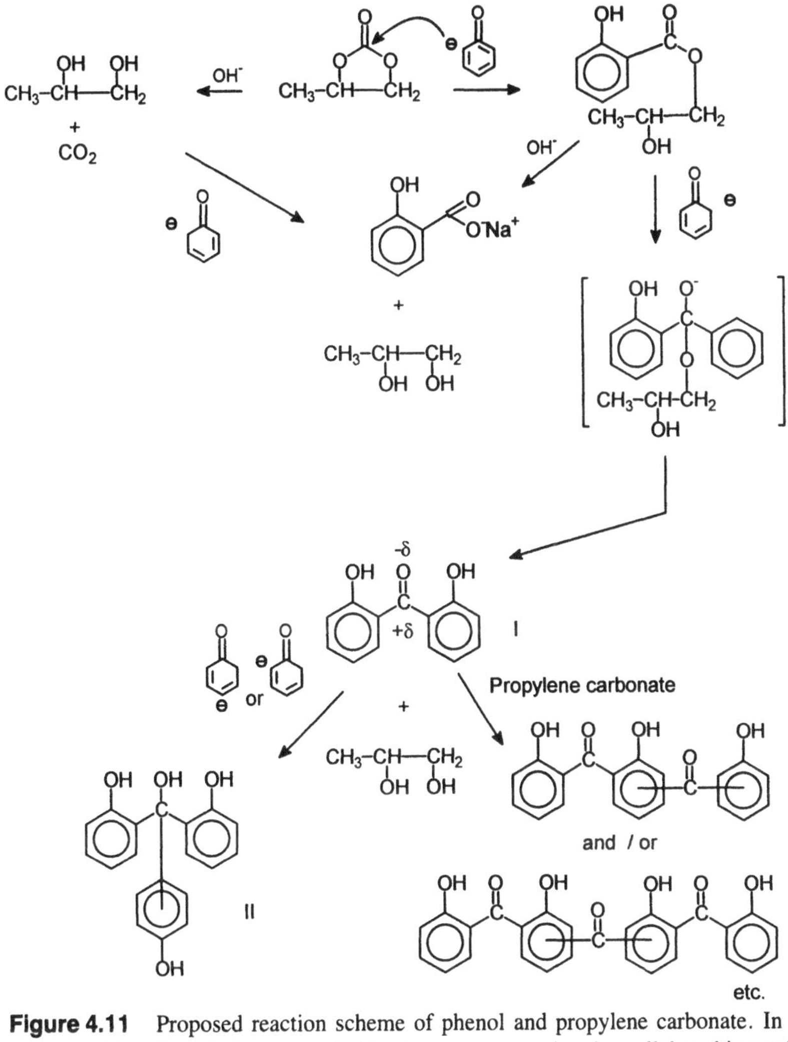

Figure 4.11 Proposed reaction scheme of phenol and propylene carbonate. In PF resins traditional methylene cross-linking is superimposed and parallel to this mechanism. (From Refs. 50 and 64.)

phenolic nucleus on which a methylol group is already present becoming from a trifunctional at least a tetrafunctional reagent (or even more functional if more than one propylene carbonate reacts), (3) by the higher concentration of reagents (the ester being a proper reagent and not a catalyst), and (4) by carbonic acid being a diprotic acid. All the causes above will induce much earlier gelling and consequently, the gelling acceleration observed. It is interesting to note that for propylene carbonate this effect starts at around pH 6 (Fig. 4.10), a pH value at which phenolic carbanions are known to exist in higher proportion than OH⁻ [1,70].

Examination of the behavior of the other esters indicates that this mechanism is also at work. Their reactions are slower than that of propylene carbonate because they function as reagents of lower functionality. Thus

and because they are monoprotic acids rather than diprotic acids such as carbonic acid. The dependence of the acceleration effect of the ester of monoprotic acids from the pK_a of the acid is not casual but is related to the intensity of the nominal $+\delta$ charge on the carbon of the carbonyl group, explaining the different intensity of acceleration.

Rather, could the nonconfirmed Na⁺-ring mechanism discussed above be responsible for the ester acceleration effect [50,54]? Unfortunately, participation of this other mechanism again cannot be relied on, due to the impossibility of confirming the existence of Na⁺ ring complexes discussed earlier [50,54]. The results obtained by using the ester acceleration mechanism for PF adhesives for particleboard under standard industrial conditions appeared to confirm positively the existence of the mechanism and its great usefulness [54]. At the slower pressing times characteristic of PF resin adhesives, the inclusion of an ester accelerator greatly improves the IB strength both dry and in the boiling test. The increasing percentage IB retention of the board is also a good indication of vastly improved durability. Even more interesting is the performance of boards pressed at faster pressing times. While a nonaccelerated industrial PF falters between 15 and 13 s mm⁻¹ press time, PF resins to which 10% triacetin ester has been added do not appear to show any significant decrease in performance, even down to 10 to 12 s mm⁻¹ pressing time [54]. Furthermore, the results are still good enough to pass exterior-grade board standards comfortably even at pressing times as fast as

7 s mm^{-1} [54]. This is clearly an exceptional result as such pressing times, 7 s mm^{-1} thickness of the panel at 190 to 200°C, are comparable to those obtained with fast UF resins under equivalent pressing conditions. It means that full exterior-grade properties can be obtained when using PF adhesives at pressing times very much faster than those achievable previously, and comparable to those of the much faster aminoplastic resins [54].

The characteristic ester fragment shown to attack the phenolic nuclei indicates that chemical compounds unrelated to esters, but which also on decomposing produced the same type of fragment, should also be capable of accelerating the curing of PF resins under alkaline conditions, by means of the same mechanism. This was found to be the case for organic acid anhydrides, in particular acetic and maleic anhydride [72]. Thus addition to the PF adhesive of 5% anhydride was found to shorten the PF gel time significantly [72]. Increasing the quantity of anhydride from 5% to 10% shortened the gel time only slightly more. Maleic anhydride appeared to accelerate PF resin curing better than did acetic anhydride. That acceleration is not due to the acid formed from anhydride hydrolysis was shown by the gel time remaining unaltered when only acid rather than anhydride was used [72]. The proposed mechanism for cure acceleration behavior appears to be the same as in the ester acceleration of PF resin curing. The attacking fragment is also likely to be the same. Thus

Organic anhydrides hydrolyze in water solutions, and PF resols are in water solution: to diminish the extent of hydrolysis, thus to diminish the proportion of the reactive fragment being transformed into acid, also further shortens the PF resin gel time. Ten percent of a surfactant added to the phenolic resin to form micelles in which some PF resin is entrapped, on addition of anhydride, which is not miscible with water, for the part of the anhydride capable of migrating within the micelles, decreases the transformation to acid. This leads to more efficient use of the mechanism on which the acceleration effect is based [72], acetic anhydride in a micellar environment shortening the PF resin gel time even further. The use of anhydrides as cure accelerators of alkaline PF resols is not practical, the use of ester accelerators being far more significant. It is of some interest, however, that compounds other than esters, able to produce the same reactive fragment, can by the same mechanism accelerate the curing of alkaline PF resins.

B. Urea Acceleration and PF Molecular Doubling

Urea can also be used as an accelerator in curing alkaline PF adhesives, the urea being added to the resin just before application to wood particles [73], or even by addition of urea to wood particles before separate addition of the PF adhesive. The reason that urea shows such behavior can be ascribed to the relative reactivities toward methylol groups of urea and phenolic nuclei under alkaline reaction conditions. A study has shown that there are definite pH ranges in which the reaction of urea, or UF resin unreacted $-NH_2$ and $-NH-$ groups, with the methylol groups carried by a phenolic resin is more favorable than is autocondensation of the PF resin [74,75] (Fig. 4.12).

A model compound study [76] using o- and p-hydroxybenzyl alcohols has shown that under alkaline conditions (pH $\simeq$ 11.5) both autocondensation of o- and p-hydroxybenzyl alcohols and condensation of the two hydroxybenzyl alcohols with urea occurred. Autocondensation of the p-isomeride appeared

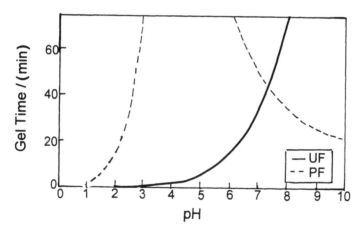

Figure 4.12 Superimposed gel-time curves as a function of pH of industrial PF and UF resins.

to be faster than that of the *o*-isomeride [76], confirming other results in an acid environment. The situation was different as regards condensation of the hydroxybenzyl alcohols with urea [76]: first, the condensation with urea of both hydroxybenzyl alcohols was more rapid than both autocondensation reactions. Second, the *o*-hydroxybenzyl alcohol appeared to condense with urea more rapidly than did the *p*-isomeride (Fig. 4.13). The presence of

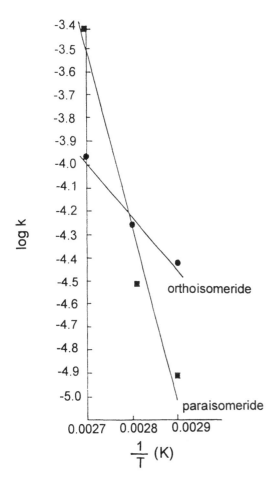

Figure 4.13 Log_{10} *I* versus 1/*T* (*T* in Relvin) for the reaction of *o*- and *p*-hydroxybenzyl alcohol with urea, detailing relative rates of condensation of *o*- and *p*-phenolic methylols with urea.

methylene bridges connecting phenolic nuclei with urea was confirmed by ^{13}C NMR [76].

Variation in physical properties of PF resins prepared using small additions of urea also appears to confirm the results obtained by model compound studies [76]: the average degree of polymerization increases when increasing amounts of urea are added, and the polymer appears to grow mostly linearly. This was confirmed by increases in viscosity, ^{13}C NMR, and GPC [76], with *o*-methylol groups on the PF resin favoring reaction with urea more than did *p*-methylol groups.

In conclusion, PF resin linear degree of polymerization values appear to be increased very rapidly by the addition of small amounts of urea to levels not easily achievable without urea. Thus

an occurrence that is supported by both the increase in viscosity and faster gel times obtained with increasing percentages of urea. This is then a method of attaining not only stable, linear, higher-molecular-weight PF adhesives, but also faster-gelling adhesives, by addition of very small amounts of urea without affecting the exterior durability of the hardened PF-bonded panel.

VII. SELF-NEUTRALIZING ACID-SET PF WOOD ADHESIVES

Acid-setting PF resins can be used to prepare adhesives capable of hardening rapidly, even at ambient temperature, and of excellent coordination, cohesion, and performance. They have not been used commercially as wood adhesives, as the very acid conditions at which they set and cure are drastic enough to cause severe hydrolysis of the structural wood constituents [77–82]. This causes severe strength loss in the wood and characteristic failures of the bonded joints, especially in the acid-weakened wood near the glue lines. This disadvantage, if overcome, could lead to:

1. The preparation of glulam and finger-joint phenolic resins in which resorcinol is not needed, thus which are considerably lower in cost and comparable in performance to PRF resins.

2. The preparation of PF thermosetting resins for particleboard and plywood capable of curing at much faster pressing times than those of the presently used alkaline PF resins (an acid-setting PF can cure as fast as 30 s at 25°C with the correct amount of catalyst).

Such resins can be used as exterior wood adhesives only if during their rapid curing an equally rapid neutralization of the glue-line pH can take place. In this manner the cured glue line does not contain free acid strong enough to damage the wood. Several neutralizing systems based on different principles were discussed by Myers [80] and other authors [83–85]. The only three promising systems presented up to now for application to thermosetting adhesives for panels were based on the use of (1) fillers or extenders impregnated with $Mg(OH_2)$, (2) mixtures of aluminum powder and Reax [80], and (3) the use of morpholine–acid complexes [83]. The two first systems were reported to give adequate initial strength and minor strength deterioration of birch plywood during high-relative-humidity, accelerated-aging tests. Unfortunately, the pH values of the cured glue lines were not reported. Thus the extent of acid neutralization, although it appears to be present as deduced from accelerated-aging tests, is not clear. The third system performance and neutralization effectiveness is better documented.

p-Toluenesulfonic acid (p-TSA) is the acid used to cure acid-setting PF resins and the curing pH values used vary between 0.2 and 1.2 for pot lives of between 1 and 3 h. It is based on the thought that to neutralize the p-TSA, a slow-release alkali or an alkali released by a particular physical condition could be used. Efficient neutralizing systems were found and described [83]. They are based on the weak acid complexes formed by the base morpholine:

The advantage of such a system is that it allows the stronger acid (the p-TSA) to set and cure the resin, then to exchange the weaker acid in the morpholine complex, with the conclusion that p-TSA (strong) is neutralized and a much weaker acid released. The final pH varied between 3 and 6 depending on which weaker acid was used in the complex. The p-TSA neutralization is started only at the trigger temperature at which the morpholine/weak acid complex decomposes. Other variations on this theme were also used in the case of thermosetting resins. Several acids weaker than p-TSA were used to form the relevant morpholine complexes. The results indicated that a neutralizing system based on complexes of morpholine is effective in neutralizing the acid contained in a hardened glue line. The curing of the resin still proceeded rapidly and the strength and wood failure results obtained in the bonded joints satisfied relevant standard specifications.

Neutralizing systems based on morpholine/phosphoric acid and morpholine/lactic acid complexes gave the glue line a final pH nearer to neutral. The morpholine/phosphoric acid complex was the one with which the best glue line strength was obtained while still obtaining a perfectly acceptable pH of the hardened glue line, higher, for instance, than is obtained with commercial acid-cured UF resin controls [83]. The results complied with the relevant weather- and boilproof wood adhesive specifications of British Standard BS 1204-1965, part 2.

The results of industrial structural exterior-grade finger joints confirmed that with radio-frequency (RF) curing, very short times are required to neutralize the residual *p*-toluenesulfonic acid present in the cured glue line. A very short curing time can also be used effectively to obtain joints suitable for structural exterior-grade applications.

Studies have shown that there are two types of self-neutralizing systems that can be used for thermosetting acid-curing PF resins.

System 1 is based on the substitution, at the temperature of decomposition of the morpholine/weak acid complex, of the p-TSA in place of the weaker acid in the complex, substitution that stabilizes upon cooling. A weaker free acid has substituted the strong free p-TSA, the system becomes less acidic, and the pH rises to acceptable levels. It is clear that in this case the weaker the acid, the higher the final pH will be. However, the weaker the acid, the greater the interference on the strength of the cured resin (due to a too-low decomposition temperature, hence premature decomposition, of the morpholine/weak acid complex). Hence the best compromise is shown by the results obtained with the morpholine/phosphoric acid complex [83].

In neutralizing system 2, based on a p-TSA/morpholine complex, the resin is maintained at an acceptably high pH, and decomposition of the complex at high temperature liberates p-TSA, which cures the resin, the glue line again self-neutralizing by re-forming the p-TSA/morpholine complex on cooling. This system causes, by its nature, slower curing of the resin [83].

In the first system several complexes of morpholine/weaker acid were tried in order to prepare a cold-setting acid-setting PF resin [83]. The results were erratic and in the cases in which a good result was achieved, the result was not reproducible. A source of heat, hence RF curing, appeared to be necessary to obtain reasonably reproducible results.

Both systems were used to prepare particleboard in the hope that the acid-setting PF resin would allow shorter pressing times than did alkaline-cured PF resins, while maintaining a nearly neutral glue line in the finished board. Although the results were encouraging for pH neutralization and reduced water absorption (due possibly to a lower pH of the bonded adhesive in the board than in alkaline PF), shorter pressing times could not be achieved under standard phenolic particleboard pressing conditions. The morpholine/weaker acid complexes were easily isolated and stored as solids, oils, or solutions.

It must be pointed out that if the acid in the cured glue line has not been neutralized, not only will the pH of the cured glue line be low but the strength of the bonded joint will also be low and wood failure percentages high. This mode of failure is characteristic of bonded joints where extensive hydrolysis of wood constituents, hence weakening of the wood, has taken place due to the acid not being neutralized.

A good joint with self-neutralizing PF resins shows high strength and high wood failure rates. Furthermore, joints cured in conditions that are far from optimum, which show high strength and low wood failure, are also a clear indication that the acid in the cured glue line has been neutralized. The level of neutralization of residual acid in the glue line appeared to be strongly dependent on temperature and time of cure.

In conclusion, these self-neutralizing PF adhesives show promise for applications where heat is used to cure the resin, thus providing the means for neutralization of the acid, such as in radio-frequency-cured exterior-grade finger joints. The advantages are mainly economic, as the elimination of resorcinol from resins used for exterior-grade structural finger joints results in about a 40% drop in the cost of resin. Self-neutralized pH values as high as 7.5 were recorded for cured PF resins immediately on cooling the joint and the relative glue line after adhesive hardening [83]. As regards the application of such a system to panel products, such as particleboard, considerably more applied work would be needed before such a system could be fully evaluated.

VIII. HYBRID PF–ISOCYANATE COPOLYMER THERMOSETTING ADHESIVES

In diisocyanate and urethane chemistry it had long been held that in mixed MDI–PF resins in water, the isocyanate group can react almost exclusively with water, this reaction being much faster than that with phenolic hydroxyls. This is correct, and consequently, diisocyanate deactivation by water was believed to ensue [1,86]. What has always been discounted is the fact that the isocyanate group ($-N=C=O$) reacts extremely rapidly with the methylol group (hydroxybenzyl alcohol group = $-CH_2OH$) in PF resols [87]. This reaction has been proved by kinetic studies to be at least an order of magnitude faster than the deactivation of polymeric MDI with water [87]. The results showed that methylols carrying PF resols can undergo two reactions of almost equal rate. One reaction is the standard condensation that PF resols undergo to cross-link through methylene ($-CH_2-$) bridges: this is in itself very slow, but when to the phenolic resol an accelerator such as urea, or particularly, a very small amount of a polyflavonoid tannin is added [87,88], it becomes very fast. The second reaction is the attack of the MDI on the hydroxyl of the methylol groups $-CH_2OH$ present on the PF resol. In such a system the reaction of deactivation of MDI with water by forming polyureas is much slower. These results imply, instead, that deactivation of MDI by water when a PF resol is present is small, and this was proven to be the case [87,88]. The mechanism involved is

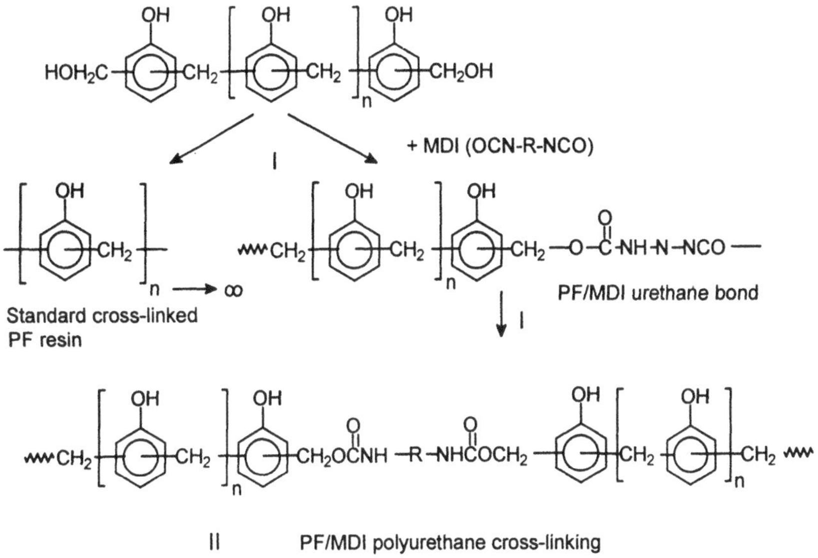

I

Standard cross-linked
PF resin

PF/MDI urethane bond

II   PF/MDI polyurethane cross-linking

Another interesting reaction, however, also appears to occur. This is the
reaction of the methylol group in a PF or UF resin with the aromatic nuclei
of nomomeric and polymeric MDI. The $-N=C=O$ isocyanate groups strongly
activate the aromatic nuclei of MDI, and the $-CH_2OH$ group of the PF resin
can react with it rapidly. Thus in the case of saligenin (o-hydroxybenzyl
alcohol) the reaction occurring appears to be

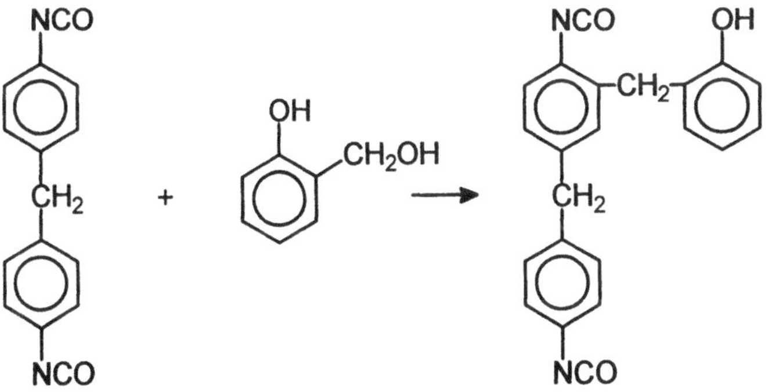

That this reaction exists can also be observed by following the reaction by IR
at different temperatures. It is particularly noticeable when the reaction is
carried out at a lower temperature (22°C).

While at 22°C this reaction is very evident due to the marked decrease in the rate of the reaction between $-CH_2OH$ and $-NCO$ groups, at 30 and 40°C the proportion of this reaction in the entire system starts to decrease markedly, at 60°C constitutes only $\pm 10\%$ of the total, and at higher temperatures disappears completely. The same can occur in the reaction of polymeric MDI with a PF resol (of which saligenin is the simplest model) or a UF resin.

In conclusion, two mechanisms exist for copolymerizing PF resins with polymeric MDI. The first is based on the reaction of the $-NCO$ group of the isocyanate with the active $-CH_2OH$ group of the PF resin. The second is the reaction of the $-CH_2OH$ group of the PF resin with the activated aromatic ring of MDI. To these can be added the well-known reaction of the PF resins to form methylene bridges. It is interesting to consider which of these two mechanisms is likely to be predominant. The two reactions of the methylol group ($-CH_2OH$) are easy to observe in the case of the reaction of saligenin with MDI. When reacting saligenin with MDI in anhydrous conditions, the two reactions can easily be followed by FT-IR and their reaction rates calculated.

Both reactions are easier to follow at 60°C, this temperature being particularly favorable in noticing the rapid disappearance of the IR band, without being so fast as to indicate that the band might not have been there in the first place [87]. The rate of both reactions increases with temperature. At higher temperatures (above 60°C) the formation of urethane bonds by reaction of the $-NCO$ with the $-CH_2OH$ of PF resins appears to predominate. At lower temperatures the formation of methylene bridges by attack of the PF resin $-CH_2OH$ on the aromatic rings of MDI appears to be the favorite. This appears to be confirmed by the relative variation in the two reaction rate constants as a function of temperature [87].

The mechanism discussed has been confirmed by the preparation of laboratory and industrial plywood panels with the MDI–PF system in water. The results obtained were excellent. At least one known plywood mill has now been in commercial operation with such a system for the last three years. The hybrid MDI–PF adhesive systems have shown remarkably good performance in the laboratory, particularly in the industrial production of marine- and exterior-grade plywood. They are very effective for difficult-to-bond wood veneers. They are quite remarkable for their adaptability to higher-moisture-content veneers. They are very easy to handle, to use, and to assemble in the factory, requiring assembly only at the glue-mixing stage, even with off-the-shelf components. Although these systems could also be used for particleboard manufacture, better and cheaper systems for this application have already been reported [89].

Important questions regarding such an adhesive system involve which would be the best type of PF resin to use, and if the tannin accelerator for the PF part of the system is truly necessary. The system can be used to good effect without the addition of the small amount of tannin accelerator [73]. A high degree of polymerization PF resin of high formaldehyde/phenol molar ratio has, for instance, been proven to yield excellent results for plywood [73], by using this system without a tannin accelerator addition.

Alternatively, the hybrid system now in use for several years is based on a low-degree-of-polymerization PF resol to which a small amount of flavonoid tannin accelerator has been added [90]. In the latter case the principles underlying the type of PF resin which should be used is that PF resols of low average degree of polymerization (DP) (thus only short-length oligomers) and a P/F ratio of 1:1.8 should be used. Thus resins with a high proportion of –CH$_2$OH on the skeletons of the short PF oligomers perform better than PF resins of higher average degree of polymerization. Instead, standard commercially available PFs have longer-length PF chains and a much lower proportion of –CH$_2$OH methylol groups. This difference is expected, as the key reaction is between the isocyanate and methylol groups. Thus the shorter the DP of the resin at an equivalent P/F molar ratio, the higher the number of –CH$_2$OH groups and the more evident the effect of urethane formation. Equally, the higher the proportion of formaldehyde to phenol, the higher the number of –CH$_2$OH groups.

It is also of considerable importance to discuss whether standard, commercial PF resins bought off the shelf could be used successfully with such adhesive systems, and what principles are involved at the molecular level in how the formulations need to be or can be changed to optimize the performance of commercial PF resins used to obtain similar results. Most commercially available PF resins work badly, or insufficiently well, with the systems and formulations presented, unless they are of the very fast type.

If a low-degree-of-polymerization PF of molar ratio 1:1.8 is not available and a commercial PF resin with tannin accelerator has to be used, there are a few ways in which the performance of MDI-based systems can be optimized according to the type of PF resin available. The main problem with commercially available PF resins is their higher degree of polymerization and relatively low level of methylol groups. What is needed, then, is to shift the overall initial degree of polymerization down and the methylol content of the resin up. The easy way to do this is to increase the percentage of tannin accelerator (and of the concomitant amount of paraformaldehyde that goes with it). Thus the tannin after the initial very fast reaction [87] with paraformaldehyde to form methylolated tannins immediately provides a phenolic component with a high degree of methylolation. An increase of a few percentage points in the tannin content of the formulation (and concomitantly, of paraformaldehyde) will easily adapt a commercial phenolic resin to a better performance of the MDI–PF system without increasing the cost (actually diminishing it). Equally, regarding both degree of polymerization and fast formation of a higher amount of methylol groups, the addition of a few percentage points of monomeric resorcinol with an increase of a few percentage points of paraformaldehyde in the cold, in the glue mix, will, by the same mechanism as for the tannin, adapt the commercial PF to the MDI–PF system. The use of resorcinol chemical is costly, however, and for this reason it is not recommended. Logically, another approach is to decrease the proportion of MDI to fit the amount of methylol groups present on the particular commercial PF resin at the end. This will optimize that particular MDI–PF system, but by lowering the amount of MDI, one weakens the formulation and might render it not good enough for difficult-to-glue hardwoods. The most appropriate system, then, is always to increase, within narrow limits, the proportion of tannin and of the amount of paraformaldehyde that goes with it. If a decrease in MDI is sought, perhaps a concomitant increase in tannin should also be sought. With these adaptations, several commercial PF resins have also given adequate results with such MDI–PF systems.

The type of polyflavonoid tannin used as accelerator also has a definite influence. Mimosa and quebracho tannins tend to give longer pot lives and slower reaction times, while pine tannin (or any other procyanidin–prodelphinidin tannins) will give shorter pot lives and faster reaction times. For correction of a commercial PF resin, the faster pine tannins will have the better effect. Mimosa tannin that has undergone a phlobatannin type of rearrangement [87] will have similar but not quite as evident results as those of pine tannin.

The type as well as the amount of tannin added to the formulation, at a parity of paraformaldehyde content, also determines the relative proportions of urethane and methylene cross-links obtained in the cured network. Thus it determines, although within fairly narrow limits, the physical properties of the cured glue line and hence its performance. This effect is due to the relative rates of the various reactions in the total system [87]. At a parity of proportions, the less reactive the tannin, the higher the proportion of urethane bonds will tend to be; the more reactive the tannin, the higher the proportion of methylene linkages [90]. There is no doubt, however, that the presence of a high proportion of a synthetic PF resin which has a much lower reactivity than tannin in the plywood formulations will shift the balance strongly in favor of urethane, rather than methylene, cross-linking. In this context the proportion of urethane cross-linking will definitely be greater than for MDI–tannin systems not containing PF resins. It is felt that it is the balance, whatever the optimum might be, of urethane and methylene cross-linking that must be varied to fit optimally veneers of a particular wood species.

In reality, the coreaction of MDI with the methylol groups of a PF resin is nothing but a method of accelerating the high-temperature curing of a PF resin. Figure 4.14 compares the difference in gel time as a function of pH of a PF resin and of a 30:70 MDI–PF system prepared using the identical PF resin. It is quite clear that the rate of curing is considerably faster with the hybrid system, which explains the excellent bonding results observed with the hybrid system. Although this system is excellent for any plywood manufacturer that has the capacity of preparing its own PF resin, or that is technologically advanced enough to be able to mix to good effect, or to adapt, a purchased PF resin with MDI, it presents a serious problem to an adhesive manufacturer. The MDI–PF mix is stable for only 4 to 5 h: an adhesive manufacturer thus cannot deliver a perfectly balanced product to a plywood mill, because the mix has too short a shelf life. This problem is overcome with ease by the use of a blocked diisocyanate. The mixture of a water-based

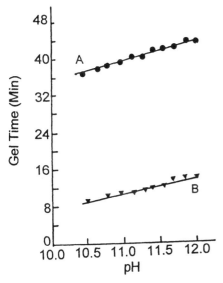

Figure 4.14 Gel times at 100°C as a function of pH of a PF resin and of the same PF resin in a 30:70 MDI–PF copolymer. (A) PF resol alone; (B) 30/70 MDI/PF resol.

PF resol with a blocked MDI is perfectly stable: it only needs the addition in the glue mix, before application, of catalytic quantities of a chemical to deblock and regenerate the isocyanate groups. Phenol itself is a good blocking agent [91], which can be used for this purpose to good effect. The phenolic resin itself, possibly supplemented with some monomeric phenol blocking agent, can be used at ambient temperature to block the isocyanate group partially [92]. The addition of small amounts of tributylamine or other amine deblocking agent in the glue mix will then regenerate the isocyanate group [91,92] and render the system operational again and again, giving excellent bonding results (Fig. 4.15).

A further refinement of this system to attain even faster high-temperature gel times, and hence even faster board pressing times, is the additional use of small amounts of a curing accelerator. Three different systems can be used. The use of flavonoid tannins for this purpose was mentioned earlier. The small amount of tannin acts as an accelerator of the autocondensation of the PF resin only. Although it promotes faster curing, this is not based on accelerated reaction of MDI–PF copolymerization. On the contrary, it subtracts methylol groups from reaction with the isocyanates. For this purpose, catalytic amounts of standard catalysts for urethane synthesis

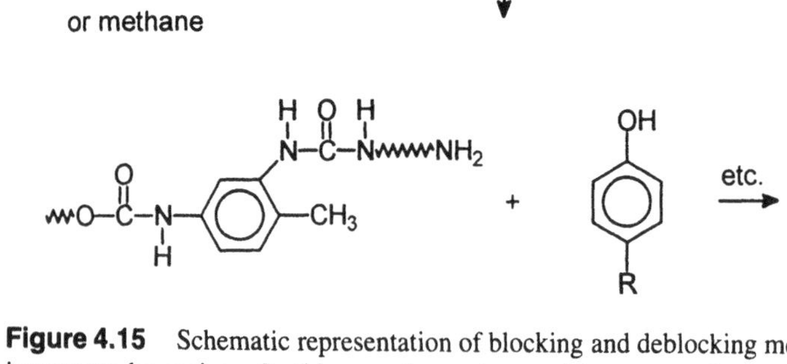

Figure 4.15 Schematic representation of blocking and deblocking mechanism of an isocyanate by amines. In this case the amine used is not a tertiary amine as in Ref. 92, but 3,3'-dimethyl-4,4'-diaminodicyclohexylmethane. (From Ref. 91.)

dispersed in the resin have been used [93,94]. Dibutyltin dilaureate and several amines have been found to accelerate consistently and considerably the reaction of PF methylols with isocyanate groups to form urethane bridges [93,94] in water-based MDI–PF adhesives. The rates of reaction determined were of an order of magnitude faster just by addition of 0.1% dibutyltin dilaureate (DBTL) on total resin solids. With 1% DBTL at 60°C the mixture gelled in just 15 min, and with 0.1% DBTL in 40 min, instead of several hours for the uncatalyzed mix [93,94]. DBTL amine and triethylamine were found to be effective [93,94]. It is evident that all sorts of other catalysts for urethane promotion [1] are likely to have an equally marked effect, and the catalysts of this type that have been mentioned here by no means exhaust the possibilities gleaned by this conceptual approach. DBTL-accelerated MDI–PF resins give excellent plywood and particleboard panels [94], and it is possible with this system to obtain very fast pressing times at temperatures much lower than those used today for panel products [94].

A third acceleration system that can be used is based on α-set types of accelerators, described earlier. These are effective not only on PF resins alone but with the hybrid MDI–PF copolymer adhesives, although to a more limited extent. It is interesting to note that with hybrid MDI–PF systems these accelerators act not only on the PF part of the adhesive in the manner already described, but in reality, also, during the course of the α-set attack, influence the rate of copolymerization by increasing the amount of hydroxybenzyl alcohol–like groups available for reaction with the isocyanate (compound II, Fig. 4.11). Kinetic acceleration of the copolymerization by increased concentration of one of the reactants also ensues. Mixing two such acceleration systems also yields faster reaction times [94].

A discussion of these hybrid MDI–PF adhesives would have little meaning without addressing the aspect of cost. For plywood a 10:90 MDI–PF system is only approximately 55 to 60% of the cost of a MF resin [90], and has costs approximately 4% higher than those of a pure PF resin [90]. If it is considered that pressing times can be as much as 50 to 60% shorter than for a pure PF resin, and comparable or faster than for MF resins, using such hybrid systems is actually much cheaper than using pure PF, pure MF, or pure MDI resins [90]. If to this is added their truly superior bond performance [90] and the possibility of bonding wood products of higher moisture content [90] and possibly even at lower temperature [93,94], such systems can readily be described as a new generation of wood adhesives with definite future prospects.

IX. PUF THERMOSETTING WOOD ADHESIVES

Phenol–urea–formaldehyde (PUF) and phenol–melamine–formaldehyde (PMUF) have acquired some importance in recent times. PMUFs, introduced to provide better water resistance than that of MUF alone, generally contain only 5 to 10% phenol. They cannot be called phenolic resins, but really behave as an upgraded MUF resin, behaving mostly like the latter [62,63]. PUFs contain a much greater proportion of phenol and have been studied for use in preparing resins of moderate water and possibly temperature resistance which, however, still involve as rapid curing and pressing times as those of UF resins.

X. PHENOL–FURANIC WOOD ADHESIVES

While phenol–furanic resins are well known in fields of application other than wood, they have not been used to any great extent in wood adhesives. There have been suggestions [98] that furfural be used as partial replacement of formaldehyde in PF resins for plywood. Their limited use can probably be ascribed to the higher cost of furfural than of formaldehyde. However, the natural origin of furfural and the pressures on formaldehyde emission may eventually lead to the use of such resins.

Baxter and Redfern [98] have proposed a procedure to prepare phenol–furfural–formaldehyde resins in which furfural was incorporated in the polymer skeleton. This was done to minimize retardation of the resin rate of cure due to the lower reactivity of furfural with phenolic nuclei in relation to the higher formaldehyde reactivity. The method proposed consisted of the preparation of a phenol–furfural resin, with furfural in molar defect, this followed by methylolation and partial further condensation with formaldehyde of the preformed phenol–furfural resin.

The use of furfural in phenolic and phenolic-like resins was also addressed in the case of phenol–resorcinol–formaldehyde and tannin–resorcinol–formaldehyde cold-set adhesives [99]. In these a system very similar to that suggested by Baxter and Redfern was used. Resorcinol-capped phenol–furfural or tannin–furfural resins were prepared but could only be used with paraformaldehyde hardener, as the reactivity of furfural even with resorcinol was too low to give the resins cold-setting adhesive capability at a reasonable rate of reaction. Phenol–furfural, resorcinol–furfural, and furfural–furfural furylmethine bridges definitely appeared to have formed.

The use of completely furanic resins, for applications similar to those for PF wood adhesives, has also been advocated. None of these resins is used at present in wood adhesives, although their performance is reported to be good, probably because their costs are higher than those for PF resins. A discussion on pure furanic resins transcends the limits of this book, and the reader is reminded that excellent reviews already exist [100]. There is no doubt, however, that with the increased awareness of the environment, furanic and phenol–furanic resin might eventually find widespread applications in wood adhesives as well [101].

XI. FORMULAS

Initial formulations of laboratory and industrial relevance are presented which might be of use to those new to the field.

A. Lower-Condensation, High-Methylol PF Resin [50,54,64,76,87,90]

Ninety-four parts by mass of phenol is mixed with 40 parts (20/80) methanol–water solution and 55 parts by mass of 96% paraformaldehyde powder. After stirring for 30 min at 40°C, the temperature is slowly, over a period of 30 min, increased to reflux (94°C). A total of 20 parts by mass of 33% sodium hydroxide solution is added in four equal parts at 15-min intervals over the total period of 60 min. Following this the mixture is kept at reflux for 60 min and then cooled. The resin solids content is 60%, the viscosity 3100 cP at 25°C, and the final pH 10.80 before adjustment to the final pH desired.

B. Preparation of PF Resin by Urea-Induced Molecular Doubling [76]

Ninety-four parts by mass of phenol is mixed with 40 parts (20:80 m/m) methanol–water and 55 parts by mass of 96% paraformaldehyde powder. After stirring for 30 min at 40°C, the temperature is slowly, over a period of 30 min, increased to reflux (94°C). A total of 20 parts by mass of 33% sodium hydroxide water solution is added in four equal parts at 15-min intervals over the total period of 60 min. Following this the mixture is kept at reflux for 60 min and then cooled. Urea is added in varying amounts (Table 4.4). For each amount a variation of the resin can be made in which the urea is added at 15, 30, or 45 min during the total refluxing period (60 min). After 60 min of refluxing, each resin is cooled. The types of gel times (at pH 11.5) and viscosities attained with variations in this basic resin are shown in Table 4.4. Many more variations are conceivable.

C. MDI–PF Glue Mixes [87,90]

To the 121 parts by mass of the liquid PF resin described first above is added at ambient temperature 10 to 30 parts by mass of raw polymeric MDI (Bayer 44V20 or VK, or similar type of any other manufacturer). The entire mixture converts instantly to a very soft, easy-flowing mass, which is homogenized easily by vigorous stirring and the addition of small amounts of water. When needed, to this is added 10 parts by mass of 45% solution of a fast procyanidin-type flavonoid tannin (pine tannin, pecan nut tannin, or gambier tannin), 1 part by mass paraformaldehyde, fillers, and extra water.

D. Higher-Condensation PF Resins

(a) [41,42] One hundred parts of phenol (by mass), in 66.8 parts water and 15.6 parts of a 40% NaOH water solution, is heated for 1 h under mechanical stirring in a reaction vessel equipped with reflux condensers and stabilized to 70°C. The required amounts of 37% formalin solution to obtain a molar ratio

Table 4.4 Gel Times and Viscosities of Identical PF Resins with Different Additions of Urea and Different Times

Urea		Gel time at 100°C (min)			Viscosity at 25°C (cP)		
Molar % on phenol	Mass %	15/45	30/30	45/15	15/45	30/30	45/15
0	0	60	60	60	300	300	300
3.30	2.1	42	47	48	1,000	840	920
5	3.2	31	41	32	1,060	960	1,060
8.3	4.3	25	36	26	3,100	1,000	2,800
13.3	8.5	14	17	24	10,000	6,500	2,950

Source: Ref. 76.

of 1.8 to 2.2, according to what is needed, is added in a number of equal aliquots, over a period of $2^{1/2}$ h, at 10-min intervals, while maintaining the reaction under mechanical stirring and at 70°C. During the last half-hour period of formalin addition the temperature of the reaction mixture is increased to 77 to 80°C. The reaction mixture is maintained at this temperature for 6 h under mechanical stirring. The reaction is halted by cooling.

E. Acid-Setting PF Resin

1. [83] An acid-setting PF resin is prepared as follows:

Phenol	514	parts by mass
Water	19	parts by mass
25% NaOH	53	parts by mass
Paraformaldehyde 96% powder	11.4	parts by mass
Paraformaldehyde 96% powder	11.0	parts by mass
Paraformaldehyde 96% powder	11.0	parts by mass
Paraformaldehyde 96% powder	11.0	parts by mass
Paraformaldehyde 96% powder	11.0	parts by mass
Paraformaldehyde 96% powder	11.0	parts by mass

Phenol, water, and 25% NaOH solution are mixed in a reflux reactor, heated to 60°C within 30 min, and refluxed for 1 h after addition of the first 11.4 parts of paraformaldehyde. The next five additions of paraformaldehyde are carried out at approximately 10-min intervals, making sure that the preceding paraformaldehyde lot has dissolved completely. The mix is heated to 80°C (within 45 min) and held at this temperature under continuous mechanical stirring for $2^{1/2}$ to 3 h. The resin is then cooled down and the pH adjusted to 6.5 to 7.0 by addition of lactic acid or 25% NaOH, as required. A characteristic glue mix of 10 g of liquid resin containing 0.5 g of methanol, 1.0 g of wood flour, and 1.1 g of 75% p-TSA has a pot life of ±55 min.

2. [83] A phenol–formaldehyde resin of 2535 cP viscosity and 55% resin solids content, pH 6.98, is best suited for acid-setting PF applications:

Phenol	34.25	parts by mass
44% formalin water solution	52.25	parts by mass
Methanol	4.15	parts by mass
32% NaOH water solution	2.45	parts by mass
65% p-TSA water solution	4.05	parts by mass
Industrial methylated spirits	2.60	parts by mass

Charge the phenol, methanol, and formalin, and cool to 40°C. Add 32% NaOH solution and check and adjust the pH to 8.8 to 9.0. Increase the temperature to 58°C, then reflux for 1 h at 58 to 62°C. Distill under vacuum 5.5 parts by mass at 58 to 62°C. Return to the reflux and maintain at reflux for 2 h. Raise the temperature to 80°C. Hold at 80 to 90°C until the water tolerance is 1:1.25 to 1:1.5, checking every 5 min. Charge two-thirds of the total amount of the 65% p-toluenesulfonic acid solution and cool to 60°C by maximum distillation. Check the pH at 60°C and adjust if necessary to 6.7 to 7.1 with p-toluenesulfonic acid solution. Vacuum distill at 54 to 58°C to remove 29 to 30 parts by mass water. Check the viscosity and distill further if necessary to obtain a viscosity of 3000 to 3200 cP at 25°C. Add industrial methylated spirits and cool below 40°C. Check the viscosity and adjust to 2250 to 2700 cP if necessary. Check the pH and adjust to 6.9 to 7.1 if necessary. Cool to 35°C and discharge. The resin's characteristics are: gel time, 80 to 170 min at 25°C (mix: resin, 200 g; hardener 65% p-toluenesulfuric acid, 12 mL); specific gravity, 1.245 to 1.253; viscosity at 25°C, 2250 to 3600 cP; water tolerance, 1:0.5 to 1:1; and pH, 6.9 to 7.5.

F. Morpholine Complexes [83]

Morpholine–acid complexes are prepared with ease by heating the components gently in concentrated water solutions to approximately 60°C. This results in a definite and fairly intense exothermic reaction: the complexes form on cooling.

REFERENCES

1. A. Pizzi, *Wood Adhesives: Chemistry and Technology*, Vol. 1 (A. Pizzi, ed.), Marcel Dekker, New York, 1983.
2. A. Pizzi and N. J. Eaton, *J. Adhes. Sci. Technol.*, *1*: 191 (1987).
3. A. Pizzi and G. De Sousa, *Chem. Phys.*, *164*: 203 (1992).
4. A. Pizzi and S. Maboka, *J. Adhes. Sci. Technol.*, *7*: 81 (1993).
5. A. Pizzi, *J. Adhes. Sci. Technol.*, *4*: 573 (1990).
6. A. Pizzi, *J. Adhes. Sci. Technol.*, *4*: 589 (1990).
7. R. Pucciariello, N. Bianchi, and R. Fusco, *Int. J. Adhes. Adhes.*, *9*: 205 (1989).
8. J. H. Freeman, *Anal. Chem.*, *24*: 2001 (1952).
9. J. J. Freeman and C. Lewis, *J. Am. Chem. Soc.*, *76*: 2080 (1954).
10. N. J. L. Megson, *Phenolic Resin Chemistry*, Butterworth, London, 1958.
11. P. D. Caesar and A. N. Sachanen, *Inst. Eng. Chem.*, *40*: 922 (1948).
12. D. A. Fraser, R. W. Hall, and A. L. J. Raum, *J. Appl. Chem.*, *7*: 676 (1957).
13. T. Yamasaki, *J. Chem. Soc.*, *56*: 307 (1953).
14. J. Reese, *Angew. Chem.*, *66*: 170 (1954).

15. E. Alvira, I. Vega, and C. Girardet, *Chem. Phys.*, *118*: 223 (1987).
16. E. Alvira, V. Delgado, J. Plata, and C. Giradet, *Chem. Phys.*, *143*: 395 (1990).
17. E. Alvira, J. Breton, J. Plata, and C. Girardet, *Chem. Phys.*, *155*: 7 (1991).
18. A. Di Nola, D. Roccatano, H. J. C. Berendsen, in *Proteins: Structure, Function and Genetics*, 1993.
19. A. Pizzi, N. J. Eaton, and M. Bariska, *Wood Sci. Technol.*, *21*: 235 (1987).
20. A. Pizzi, M. Bariska, and N. J. Eaton, *Wood Sci. Technol.*, *21*: 317 (1987).
21. S. Proszyk and R. Zakrzewski, Activation energy of curing reaction of phenolic resins in the presence of some selected species of wood, unpublished, 1986.
22. R. Zakrzewski and S. Proszyk, *Rocz. Akad. Roln. Poznaniu* (*Ann. Acad. Agric. Poznan*), *117*: 91 (1979).
23. H. Mizumachi, *Wood Sci.*, *6*(1): 14 (1973).
24. H. Mizumachi and H. Morita, *Wood Sci.*, *73*(3): 256 (1975).
25. S.-Z. Chow, *Wood Sci.*, *1*(4): 215 (1969).
26. M. V. Ramiah and G. E. Troughton, *Wood Sci.*, *3*(2): 120 (1970).
27. B. A. Kottes Andrews, R. M. Reinhardt, J. G. Frick, and N. R. Bertoniere, Chapter 5 in *Formaldehyde Release from Wood Products* (B. Meyer, B. A. Kottes Andrews, and R. M. Reinhardt, eds.), ACS Symposium Series No. 316, American Chemical Society, Washington, D.C., 1986.
28. M. Cherubim and F. Henn, *Mitt. Dtsch. Ges. Holzforsch.*, *57*: 165 (1971).
29. S.-Z. Chow and H. N. Mukai, *Wood Sci.*, *4*(4): 202 (1972).
30. F. Mora, F. Pla, and A. Gandini, *Angew. Makromol. Chem.*, *173*: 137 (1989).
31. W. E. Johns, Chapter 3 in *Wood Adhesives: Chemistry and Technology*, Vol. 2 (A. Pizzi, ed.), Marcel Dekker, New York, 1989.
32. A. J. Kinloch, *Adhesion and Adhesives: Science and Technology*, Chapman & Hall, London, 1987.
33. B. Mtsweni, M.Sc. thesis, University of the Witwatersrand, Johannesburg, South Africa, 1994.
34. B. Mtsweni, A. Pizzi, and W. Parsons, *J. Appl. Polym. Sci.*, in press (1994).
35. M. E. Brown, *Introduction to Thermal Analysis*, Chapman & Hall, London, 1988.
36. H. E. Kissinger, *Anal. Chem.*, *29*(11): 1702 (1957).
37. G. Myers, in *Wood Adhesives in 1985: Status and Needs*, Forest Products Research Society, Madison, Wis., 1986.
38. H. Poblete and E. Roffael, *Holz Roh Werkst.*, *43*: 57 (1985).
39. G. G. Allan and A. N. Neogi, *J. Adhes.*, *3*: 13 (1971).
40. G. C. Bond, *Heterogeneous Catalysis: Principles and Application*, Oxford Science Publishers, Clarendon Press, Oxford, 1987.
41. L. Panamgama, M.Sc. thesis, University of the Witwatersrand, Johannesburg, South Africa, 1993.
42. L. Panamgama and A. Pizzi, *J. Appl. Polym. Sci.*, in press (1993).
43. D. D. Werstler, *Polymer*, *27*: 750 (1986).
44. S. A. Sojka, R. A. Wolfe, E. A. Dietz, B. F. Dannels, *Macromolecules*, *12*(4): 767 (1979).

45. L. Panamgama, unpublished results, 1993.
46. E. E. Ferg, A. Pizzi, and D. Levendis, *J. Appl. Polym. Sci.*, *50*: 907 (1993).
47. B. Tomita and H. Hatono, *J. Polym. Sci. Chem. Ed.*, *16*: 2509 (1978).
48. R. Marutzky, Chapter 10 in *Wood Adhesives: Chemistry and Technology*, Vol. 2 (A. Pizzi, ed.), Marcel Dekker, New York, 1989.
49. B. P. Barth, Chapter 23 in *Handbook of Adhesives*, 2nd ed. (I. Skeist, ed.), Van Nostrand Reinhold, New York, 1977.
50. A. Pizzi and A. Stephanou, *J. Appl. Polym. Sci.*, *49*: 2157 (1993).
51. D. A. Fraser, R. W. Hall, P. A. Jenkins, and A. L. J. Raum, *J. Appl. Chem.*, *7*: 701 (1957).
52. A. Pizzi, *J. Appl. Polym. Sci.*, *24*: 1247 (1979); *J. Polym. Sci. Polym. Lett.*, *17*: 489 (1979).
53. A. Pizzi and P. Van der Spuy, *J. Polym. Sci. Chem, Ed.*, *18*: 3447 (1980).
54. A. Pizzi and A. Stephanou, *Holzforschung*, *48*: 35 (1994).
55. J. H. Freeman, *Anal. Chem.*, *24*: 955 (1952).
56. J. H. Freeman, *J. Am. Chem. Soc.*, *74*: 6257 (1952).
57. M. Seto and K. Ozaki, *J. Chem. Soc. Jpn.*, *56*: 936 (1953).
58. S. R. Finn and J. W. James, *Chem. Ind.*, 1253 (1954).
59. S. R. Finn and J. W. James, *J. Appl. Chem.*, *6*: 466 (1956).
60. S. R. Finn and J. W. G. Musty, *J. Soc. Chem. Ind. London*, *69*, Suppl. Issue 1: S3 (1950).
61. H. L. Bender, *Mod. Plast.*, *30*: 136 (1953); *31*: 115 (1954).
62. J. Van Niekerk, *Particleboard Symposium Proceedings*, Washington State University, Pullman, Wash., 1993.
63. J. Van Niekerk and A. Pizzi, *Holz Roh Werkst.*, in press (1993).
64. A. Pizzi and A. Stephanou, *Holzforschung*, *48*: 150 (1994).
65. G. W. Westwood and R. Higgins, British patent GB2 158448A (1985).
66. Borden Inc., Japan Kokai Tokkyo Koho, J.P. 1-132650A (1989); U.S. patent priority 87-102665 (1987).
67. P. H. R. B. Lemon, *Inst. J. Mater. Prod. Technol.*, *5*(1): 25 (1990).
68. H. S. Lilley and D. W. J. Osmond, *J. Soc. Chem. Ind. London*, *66*: 425 (1947).
69. W. J. Brittain and J. B. Dicker, *Macromolecules*, *22*: 1054 (1989).
70. P. Sykes, *A Guidebook to Mechanisms in Organic Chemistry*, Longman, London, 1963.
71. R. T. Morrison and R. N. Boyd, *Chimica Organica*, Ambrosiana, Milan, Italy, 1965.
72. A. Pizzi and A. Stephanou, *J. Appl. Polym. Sci.*, *51*: 1351 (1994).
73. Bakelite A. G., personal communication, 1993.
74. N. Meikleham, Scanning and preparation of foams and foundry core binders from natural and synthetic phenolics, M.Sc. thesis, University of the Witwatersrand, Johannesburg, South Africa, 1993.
75. N. Meikleham and A. Pizzi, *J. Appl. Polym. Sci.*, in press (1993).
76. A. Pizzi, A. Stephanou, I. Antunes, and G. De Beer, *J. Appl. Polym Sci.*, *50*: 2201 (1993).

77. K. Egner, *Holz Zentralbl.*, *135*: 1857 (1952).
78. A. Müller, *Holz Roh Werkst.*, *11*: 429 (1953).
79. E. Plath, *Holz Roh Werkst.*, *11*: 466 (1953).
80. G. E. Myers, *For. Prod. J.*, *33*: 49 (1983).
81. J. S. Sodhi, *Holz Roh Werkst.*, *15*: 261 (1957).
82. F. A. Cameron and A. Pizzi, *J. Appl. Polym. Sci. Appl. Polym. Symp.*, *40* (1983).
83. A. Pizzi, R. Vosloo, F. A. Cameron, and E. Orovan, *Holz Roh Werkst.*, *44*: 229 (1986).
84. M. Higuchi and I. Sakata, *Mozukai Gakkaishi*, *25*: 496 (1979).
85. G. T. Tiedeman and M. F. Gillern, U.S. patent 3,872,051 (1975).
86. G. W. Ball and R. P. Redman, 1978, *Proceedings of the FESYP International Particleboard Symposium*, Hamburg, 1978, pp. 121–126.
87. A. Pizzi and T. Walton, *Holzforschung*, *46*: 541 (1992).
88. A. Pizzi, *Holz Roh Werkst.*, *40*: 293 (1982).
89. A. Pizzi, E. P. Von Leyser, J. Valenzuela, and J. Clark, *Holzforschung*, *47*: 168 (1993).
90. A. Pizzi, J. Valenzuela, and C. Westermeyer, *Holzforschung*, *47*: 68 (1993).
91. K.-H. Hentschel, E. Jurgens, and W. Wellner, *54th Conference of German Society of Chemists, Working Group on Coatings and Pigments*, Bad Kissingen, Germany, Sept. 8–10, 1987.
92. J. M. Zhuang and P. R. Steiner, *Holzforschung*, *47*(5): 361 (1993).
93. R. Krause, B.Sc.(Hons.) project report, University of the Witwatersrand, Johannesburg, South Africa, 1993.
94. P. Cheesman, B.Sc.(Hons.) project report, University of the Witwatersrand, Johannesburg, South Africa, 1993.
95. B. Tomita, *Adhesives for Tropical Woods*, Taipei, Taiwan, 1992.
96. B. Tomita and C.-Y. Hse, *J. Polym. Sci.*, in press (1993).
97. B. Tomita and C.-Y. Hse, *J. Polym. Sci.*, in press (1993).
98. G. F. Baxter and D. V. Redfern, U.S. patent 2,861,977 (1958).
99. A. Pizzi, E. Orovan and F. A. Cameron, *Holz Roh Werkst.*, *42*: 12 (1984).
100. W. J. McKillip, Chapter 29 in *Adhesives from Renewable Resources* (R. W. Hemingway and A. H. Conner, eds.), ACS Symposium Series No. 385, American Chemical Society, Washington, D.C., 1989.
101. W. J. McKillip, private communication, 1993.

5
Tannin-Based Wood Adhesives

I. INTRODUCTION

Condensed or polyflavonoid tannin–formaldehyde wood adhesives have been used industrially and successfully since the very early 1970s [1] for exterior wood bonding of products such as particleboard, plywood, glulam, and finger jointing. Comprehensive reviews [1–3] exist describing both the characteristics of flavonoid tannins for adhesives and their formulation into adhesives. It is not possible in a short introduction to summarize all the relevant literature on tannin adhesives and their inception; for this the reader is directed to the relevant reviews [1–3] and also to the earlier, more specialized literature. In this chapter we concentrate on much later additions to the literature that have acquired or are likely to acquire significance for wood adhesive applications.

Tannin–formaldehyde adhesives are obtained by hardening of polymeric flavonoids of natural origin, or condensed tannins, by polycondensation with formaldehyde. As tannins are phenolic in nature, one would imagine that the standard technology used to prepare and use synthetic PF adhesives would apply. This is not the case, however; the phenolic nuclei in tannins do react with formaldehyde, and with this the similarity ends. The much higher reactivity of tannins toward formaldehyde due to their A-ring resorcinolic or phloroglucinolic nuclei ensures rates of reactions, under parity of conditions, which are between 10 and 50 times faster than the reaction of phenol with formaldehyde. This denies the formation of tannin–resols, tannin resins carrying methylol (–CH$_2$OH) reactive groups, as these will condense with other tannin phenolic nuclei in a very short time. Thus resol resins, which dominate synthetic PF resin technology, are not a feasible alternative with tannins. In short, tannin resols are not stable and their shelf life is far too short to be of industrial significance. This leads to tannins being added to by a hardener, generally paraformaldehyde, but also hexamine and urea–HCHO concentrates, in the glue mix before application to wood. Thus unless a hardener is added, the tannin resin is inactive and has an indefinitely long shelf life in both liquid and powder (spray-dried) form. Pot life and shelf life are very different in TF adhesives, not identical as in the case of PF resins. Second, it is not necessary to prebuild the polymer as in the reaction of phenols with formaldehyde: the tannin extract itself is composed predominantly of a variety of polymeric oligomers which constitute the polymer. This leads to low amounts of formaldehyde being required, only for hardening. Their fast reactivity also leads to much faster gel times and hence faster pressing times than for synthetic PF adhesives, and to truly exceptional exterior-grade performance.

There are some fundamental differences between different polyflavonoid tannins that can be used for thermosetting wood adhesives. Flavonoid units in such tannins present phloroglucinol or resorcinol A-rings and catechol or pyrogallol B-rings.

The repeating units are linked to each other C4–C6 or C4–C8, the former predominating in tannins composed primarily of fisetinidin (resorcinol A-ring, catechol B-ring) and robinetinidin (resorcinol A-ring, catechol B-ring) repeating units. The C4–C8 interflavonoid linkage predominates in tannins composed of catechin (phloroglucinol A-ring, catechol B-ring) and gallocatechin (phloroglucinol A-ring, pyrogallol B-ring) repeating units. When the polymeric tannins are composed of fisetinidin and robinetinidin units, the polymers are called profisetinidin and prorobinetinidin, respectively; when they are composed of catechin and gallocatechin, the polymers are called procyanidin and prodelphinidin, respectively. The free C6 and/or C8 sites on the A-ring are the sites reactive with formaldehyde, due to their strong nucleophilicity to form adhesives under the usual conditions under which these materials are used.

Procyanidins favor cleavage of the interflavonoid linkage rather than heterocycle opening [3]. Prodelphinidin tannins show cleavage of the interflavonoid bond, but due to the effect of the pyrogallol B-ring can also present easy cleavage of the heterocycle pyran ring [4,5]. There are recent indications that the second reaction is, under certain conditions, favored and that this is of significance in, for instance, pecan tannin wood adhesives. Profisetinidins also appear to behave as prorobinetinidins, but elements of procyanidin behavior for acid and alkaline cleavage also appear to be present [4], at least in natural polymeric tannins.

Simple model compound reaction work has for many years been the staple on which knowledge of the behavior of the more complex natural tannins has been based. Although very valuable, this work has recently been shown not always to represent what the situation is as to the natural polymeric tannins, especially when these need to be modified for use in adhesives. Two reactions advanced as being particularly important for wood adhesives by model compound work have been shown instead, to be of rather limited significance for wood adhesion. The first is the reaction observed on dimers leading to phlobatannins in acid and alkaline environments [6]. The model work indicated the following:

Supposedly, "liberation" of the resorcinol A-ring leads to improved adhesives. Although there is no doubt that this reaction also occurs, but to a limited extent [7,8], in natural tannins, the extent to which it occurs is such that only some acceleration of the reaction of the liberated resorcinol nuclei with formaldehyde results [7,8]. The effect is noticeable only in lower-degree-of-polymerization profisetinidin and prorobinetinidin tannins, the tannins that react most slowly with formaldehyde [8]. No increase in number of cross-linking sites is obtained, and thus the "ultimate" strength of the adhesive (at long press and cure times) is unaffected. The increased strength noted at faster pressing times for thermosetting adhesives is due only to the increased reactivity toward HCHO of the liberated resorcinol A-ring: the same effect is obtainable simply by increasing the pH of the tannin solution, at ambient temperature, as shown later in this chapter, without the need for expensive and unnecessary modification [7]. It is of some use for the slower tannins in their application as cold-setting adhesives to diminish the amount of resorcinol added. Such a modification does not occur in procyanidin and prodelphinidin tannins (hence with the tannins which are faster reacting with formaldehyde).

The second finding of interest, a positive one, is the fact that the favorite cleavage of procyanidins is at the interflavonoid bond. Hemingway et al. [3] have used this approach to good effect to prepare cold-set glulam adhesives based on procyanidin and prodelphinidin tannins in which the added resorcinol chemical is linked directly to the C4 of the tannin.

This is a clever, and useful modification indeed, except for two facts. First, rearrangements of the catechnic acid type appear to occur in the flavonoid resorcinol adduct (which is the cold-set adhesive intermediate) [9] with corresponding loss of cross-linking sites. This does not appear to detract from its use, as positive applied results have been reported [10] with this approach.

Second, why use such a modification when separate-application fast-set "honeymoon"–type adhesives provide equally good results for gluelam and finger jointing, when also using procyanidin and prodelphinidin tannins, without modification of the tannin except as to the pH [11]? This easier approach has also proved successful on an industrial level in Europe [12] and is in current use in two countries, one in Europe and one in South America, with a procyanidin (the latter) and a prodelphinidin (the former) tannin.

The third model compound reaction of recent times of interest to tannin adhesives is the catechinic acid rearrangement [13]:

(2) R = CH$_2$

This has been advanced as a negative rearrangement that might well influence performance of the fast-reacting tannins, such as pine, pecan, and gambier tannins. Such a modification is, however, easily and inexpensively control-lable, and thus excellent adhesives, both thermosetting [14] and cold-setting [3,10], have been prepared with procyanidin and prodelphinidin tannins. In particular, thermosetting adhesives based on procyanidin and prodelphinidin tannins have been shown capable of performing considerably better than the slower flavonoid tannins on which, traditionally, tannin adhesives have been based.

The technology presented in the remainder of this chapter is based on the results obtained with adhesives based on five commercial tannin extracts: mimosa (*Acacia mearnsii*, or *mollissima*) bark tannin, quebracho (*Schinopsis balansae*) wood tannin, pine (*Pinus radiata*) bark tannin, pecan (*Carya illinoensis*) nut pith tannin, and gambier (*Uncaria gambir*) shoot and leaf tannin.

II. TANNIN-CELLULOSE INTERFACE AT THE MOLECULAR LEVEL

The interaction energies obtained from both constrained and unconstrained force-field computation of monoflavonoids and biflavonoids with cellulose tend to confirm what has already been described for synthetic PF resins. There is only one known molecular mechanics study on flavonoid–cellulose surface interactions [15]. In this, the interaction of (+)-catechin and of the biflavonoid 4,6-*cis*,*trans*-(+)-fisetinidol/4,6-*cis*,*trans*-(+)-fisetinidol with crystalline cellulose I was studied [15]. Both a constrained force-field program [16] and an unconstrained force-field program [16] were used, with coincident results. To determine the nonbonded interaction energies between the adsorbed molecules and the substrate, the steric energies of the individual isolated molecules had to be subtracted from the total steric energies of the combined systems [e.g., (+)-catechin adsorbed on crystalline cellulose]. It was found that (+)-catechin favored being adsorbed onto crystalline cellulose by a value of 31.28 kcal mol^{-1}, while the 4,6-*cis*,*trans*-(+)-fisetinidol/4,6-*cis*,*trans*-(+)-fisetinidol biflavonoid flavored adsorption by cellulose by 13.59 kcal mol^{-1}. Observation of the final geometry of the biflavonoid revealed why its interaction energy with cellulose is weaker than that of monomeric catechin. The (+)-catechin at the minimum energy conformation of the entire system lay almost flat across the surface formed by cellulose chains and interacts strongly with both chains. On the other hand, the 4,6-linked biflavonoid is seen to interact with only one of the two surface cellulose chains of the model, and this interaction is limited to a small portion of the biflavonoid molecule. This geometry explains the differences obtained. It is not possible, of course, to extrapolate the finding with a single flavonoid dimer to other flavonoid dimers or higher oligomers, the interaction of which with cellulose can be much stronger or much weaker than for 4,6-*cis*,*trans*-(+)fisetinidol/4,6-*cis*,*trans*-(+)-fisetinidol. It is important to note, however, that interaction exists, that it is considerable, and that it can be calculated. More useful, and more representative considering the innumerable isomerides of biflavonoids and higher oligomers, are the results for (+)-catechin monomer. An attractive interaction in excess of 31 kcal mol^{-1} is considerable and can effectively explain weakening of the heterocycle ether bond leading to accelerated and easier opening of the pyran ring in a flavonoid unit, as well as the facility with which hardening by autocondensation (described later) can occur. As in synthetic PF resins, the same effect explains the decrease in energy of activation of the condensation of polyflavonoids with formaldehyde, leading to exterior wood adhesives [72].

III. COMPARATIVE ^{13}C NMR OF TANNIN EXTRACTS

^{13}C NMR analysis can indicate rapidly, simply, and directly on concentrated solutions of the tannin extract, structural characteristics and properties that are of potential importance in the use of these materials as wood adhesives [17,18]. A recent study investigated by ^{13}C NMR the characteristics of five types of commercial and industrial tannin extracts [17] usable for wood adhesives: (1) black wattle or mimosa (*Acacia mearnsii*, formerly *mollissima*, de Wildt) bark extract, (2) quebracho (*Schinopsis balansae*, variety *chaqueno*) wood extract, (3) pine (*Pinus radiata*) bark extract, (4) pecan (*Carya illinoensis*) nut pith extract, and (5) gambier (*Uncaria gambir*) leaf and shoot extract. Properties of relevance found to vary considerably among the five tannins that could also be detected by ^{13}C NMR were (1) the relative proportions of fisetinidin–robinetidin units in relation to catechin–epicatechin units, (2) differences in the number-average degree of polymerization, (3) the extent of open heterocycle forms present, (4) the extent of proper branching, and (5) the relative proportion of pyrogallol versus catechol B-rings in the flavonoid repeating units. As knowledge of such characteristics are fundamental to tannin extract application and formulation as adhesives, it is of interest to discuss the analysis in some detail. Such an approach is of effective use not only in understanding the five tannins mentioned, but also in understanding any other suitable tannin that might be considered for adhesives in the future. First, it is the trends in relative ^{13}C NMR band intensities that are characteristic of different carbons in polyflavonoids which allow us to identify the significant differences among extracts as regards their use as tannin–formaldehyde adhesives.

The relative intensities of the free (unreacted) C6 and C8 sites on A-rings (Table 5.1) (Figs. 5.1 to 5.6) at 96 to 98 and 95 to 96 ppm, respectively, are a set of very sensitive bands indicating the reactivity directly and the degree of polymerization of the tannin indirectly. In this regard the five tannin extracts can be divided into two classes [17]: (1) mimosa and quebracho, which present much lower intensity of these two ^{13}C NMR bands (Figs. 5.2 and 5.3; Table

Table 5.1 Comparative ^{13}C NMR Band Assignment and Relative Band Intensities (%) for Pure Catechin and Five Types of Polyflavonoid Tannin Extracts Suitable for Adhesive Preparation

	C5,C7 (156–157)	C9 (155)	C3',C4' (145–146)	C1 (131)	C6* (120–121)	C5',C2'* (115–117)	Phloroglucinol interflavonoid (110)	C10 (101)	C6 (96–98)	C8 (95–96)	C2 (81–82)	C3 (67–68)	Catechinic acid (31–32)	C4 (27–28)
Pure catechin	97,100	58	72,76	73	97	98,96	—	68	88	97	92	93	—	80
Mimosa bark	53	30	100	44	25	26	51	19	21	31	—	—	—	21
Quebracho wood	63	33	100	66	56	99	40,52	25	20	20	—	—	—	32
Pine bark	71	71	100	49	39	76	60	37	47	23	—	—	—	37
Pecan nut pith	66	70	100	62	30	33	66	42	40	29	—	—	79	37
Gambier	100	50	100	55	69	106,95	27	50	60	28	—	—	61	61

Source: Ref. 17.

157

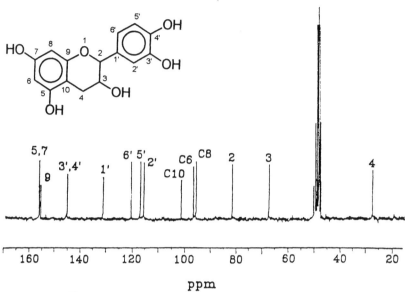

Figure 5.1 ^{13}C NMR spectrum of pure monomeric catechin with carbon atom assignments. (From Ref. 17.)

5.1); and (2) the set composed of pine, pecan nut, and gambier, in which these two bands have much greater intensities. Within each of these two sets, clear differences between the tannins can be seen. Quebracho, for instance, presents a total free C6 + C8 band intensity much lower than that of mimosa, indicating a probable higher degree of polymerization of the former (Table 5.1; Figs. 5.2 and 5.3). Furthermore, it presents a much lower intensity of the free C8 band, implying a higher proportion of C4–C8 interflavonoid linkages than in mimosa, in which the proportion of C4–C8 linkages is known to be relatively low [19]. Thus, while mimosa units are predominantly C4–C6 linked with no more than 13 to 15% C4–C8 linkages, quebracho, while still presenting a predominant amount of C4–C6 linkages, shows clearly by inference a much higher proportion of C4–C8 linkages (about 20%) than mimosa: these facts have been ascertained by other techniques [20]. The higher degree of polymerization and higher proportion of C4–C8 linkages imply that quebracho is a very "branched" tannin.

In the second set of tannins, the high proportion of free C6 clearly confirms the predominance of the C4–C8 interflavonoid linkage in these procyanidin- and prodelphinidin-based tannins, with pine and pecan nut showing a very similar degree of polymerization. The high value of the free C6 and C8 band intensities indicates that gambier tannin has, instead, a much lower degree of

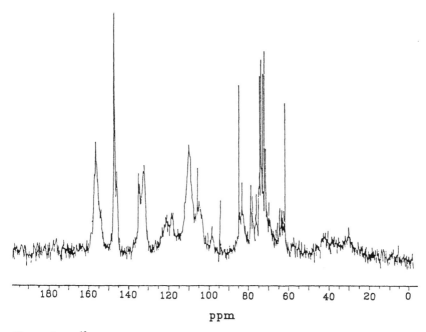

Figure 5.2 ^{13}C NMR spectrum of industrial mimosa bark tannin extract. (From Ref. 17.)

polymerization. Pure catechin, completely monomeric, presents the highest values of these two bands (Fig. 5.1). The low value for the number-average degree of polymerization (DP_n) for gambier tannin leads to the highest proportion of free C8, given by the free C8-band relative intensity (Table 5.1). Small differences are also notable between pecan nut and pine tannin: the former has a slightly higher intensity of the free C8 band and much lower intensity of the free C6 band, hence a higher number of free C8 sites, implying a noticeably higher proportion of C4–C6 linkages than in pine tannin (Figs. 5.5 and 5.6).

Confirming evidence of the above can be gathered from the conspicuous but less sensitive band at 110 ppm, indicating C8 and C6 sites participating in C4–C8 and C4–C6 interflavonoid linkages, in procyanidin–prodelphinidin tannins. The low relative intensity of this band confirms, for instance, the very low $\overline{DP}_n$ value of gambier tannin (Fig. 5.6; Table 5.1); its high intensity confirms that pecan nut tannin has one of the highest $\overline{DP}_n$ values, followed fairly closely by pine tannin. Again, the low relative intensity of this band indicates that quebracho and mimosa tannins appear to be much less polymerized, at least

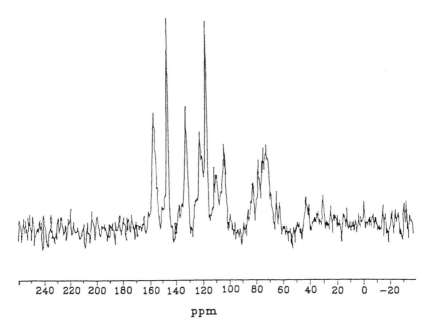

Figure 5.3 ^{13}C NMR spectrum of industrial quebracho wood tannin extract. (From Ref. 17.)

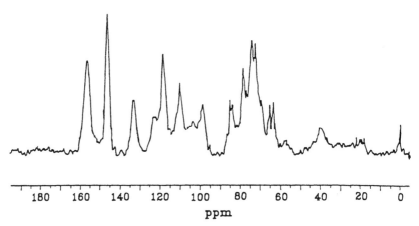

Figure 5.4 ^{13}C NMR spectrum of industrial pine bark tannin extract. (From Ref. 17.)

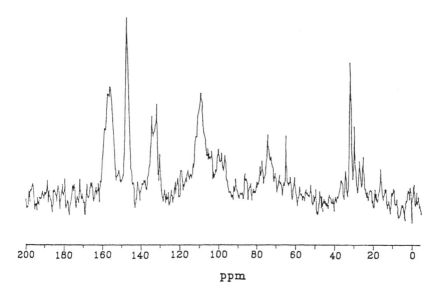

Figure 5.5 ^{13}C NMR spectrum of industrial pecan nut pith tannin extract; note the extensive $-CH_2-$ from noncompleted catechinic acid rearrangement around 30 to 40 ppm ($-CH_2-$ in DEPT ^{13}C NMR spectrum). (From Ref. 17.)

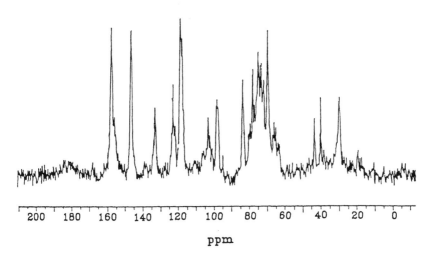

Figure 5.6 ^{13}C NMR spectrum of industrial gambier shoots tannin extract; note the extensive $-CH_2-$ from catechinic acid around 30 to 40 ppm confirmed by DEPT spectrum. (From Ref. 17.)

through phloroglucinol A-ring sites (they contain much lower amounts of phloroglucinol A-rings, thus much lower amounts of procyanidin or prodelphinidin units), with quebracho having slightly higher DP values or a higher proportion of procyanidin–prodelphinidin units than mimosa. The free C4 band (at 27 to 28 ppm) intensity is also in line with all of the above. The relative relation of their degree of polymerization is confirmed by the relative proportion of the typical (not average) viscosities of tannin extracts presented in Table 5.2. Some differences in the process of industrial extraction occurs: for "natural" quebracho, 5% metabisulfite and 97 to 100° C are used for extraction; 5% at 70° C is used in pecan; 2.5% and 70° C in pine; and no metabisulfite and 97 to 100° C for mimosa. Use of metabisulfite, for instance, will also tend to extract some higher-molecular-mass tannins to a certain extent, increasing molecular mass by autocondensation [21], while the higher temperatures tend to increase slightly the amount of carbohydrates extracted. While such differences in extraction procedures have been shown to influence the final viscosity of the extract, these do not explain differences as extensive as those shown in Table 5.2. These must be attributed in high proportion to the structure and degree of polymerization of the tannin in each extract. The viscosities reported in Table 5.2 are consistent with the qualitative relative order of the average DP value of the tannin extracts, which can be deduced by comparing the relative intensities of the relevant ^{13}C NMR bands discussed above and reported in Table 5.1. The correlation of relative ^{13}C NMR band intensities and of typical viscosities in Tables 5.1 and 5.2 is clearly nonlinear but does not follow a simple nonlinear relationship.

Table 5.2 Comparative Viscosities of Various Tannin Solutions at Different Concentrations at 25°C (Pa·s)

	Extract solids content (% by mass)		
	50	47	40
Mimosa bark tannin extract	1.8	—	0.25
Quebracho wood tannin extract	14.2	7.7	0.57
Pine bark tannin extract	—	4.4	0.51
Pecan nut pith tannin extract	57.0	—	1.8
Gambier tannin extract	3.0	1.7	0.32

Source: Ref. 17.

Linked with the above are the intensities of the C5 and C7 bands (aromatic >C–OH) at 156 to 157 ppm, generally appearing as a single broad band (Table 5.1; Figs. 5.1 to 5.6). This is a reasonably sensitive band as regards intensity. It also makes possible a qualitative indication of the proportion of phloroglucinol A-rings, thus of procyanidin–prodelphinidin units, in each tannin extract. Again confirming the other evidence, pure catechin and gambier tannin register the highest band intensity and have comparable values (Table 5.1), as seen clearly when comparing Figs. 5.1 and 5.6 with the other figures, indicating that gambier is completely catechinic in nature and again confirming its very low average DP_n value. Again two sets of tannins are evident: (1) mimosa and quebracho and (2) the other three tannins. Mimosa and quebracho polyflavonoids presenting C5 not linked to –OH groups are in relative proportions, as observed by other techniques [19]. Thus to their mainly profisetinidin–prorobinetinidin nature are super-imposed procyanidin–prodelphinidin units in 13 to 15% proportion for mimosa and 18 to 20% for quebracho. If one considers, then, gambier tannin as composed of completely catechinic structures, pine tannin from the relative intensity of this band should present approximately 10 to 15% profisetinidin–prorobinetinidin units (with resorcinol A-rings) (cf. Figures 5.6 and 5.4; Table 5.1). In pecan nut tannins the proportion of procyanidin units appears to be lower and of profisetinidin units noticeably higher than in pine tannins (a fact supported by the lower free C6 band in pecan) (Fig. 5.5; Table 5.1).

All the above leads to the conclusion that ^{13}C NMR analysis also predicts the reactivity of formaldehyde with tannins to be higher the higher the proportion of units with phloroglucinol A-rings presenting free reactive sites. This is well proven experimentally. It is, however, interesting to see within each of the two sets of tannins why there are differences in their gel times. Pine and pecan nut tannins give gel times in line with their relative average DP_n value under identical reaction conditions (Table 5.3). Although mostly catechinic, gambier gives much slower gel times, the faster reaction rate ascribed to the higher phloroglucinol A-ring content and higher proportion of free C6 and C8 reactive sites being strongly counterbalanced by the slowing down of the reaction due to the much greater number of steps needed to gelation as a consequence of the low $\overline{DP}_n$ value. For quebracho and mimosa, gel times are somewhat comparable (Table 5.3), a clear indication that as already established for mimosa [1], the phloroglucinol A-rings in quebracho appear to be mostly in angular tannin configuration and are not likely to contribute much to the rate of reaction. The higher quebracho average $\overline{DP}_n$ value indicates that in viscous solution, as used in wood adhesive applications,

Table 5.3 Comparative Gel Times at 94°C of Tannin Extracts–Formaldehyde for 40% Aqueous Solution of the Tannin Extracts

pH	Gel time (s)				
	Pecan nut	Pine	Gambier	Quebracho	Mimosa
4.8	51	86 (natural) pH	217 (natural) pH	—	516
5.0	49	—	194	—	—
5.38	42 (natural) pH	—	—	—	—
5.5	41	74	146	534	180
6.0	37	60	113	338	165
6.5	33	48	89	214	—
7.0	28	39	72	136	102
7.5	25	32	58	86	82
8.0	22	26	48	54	59
r^a	0.98	0.95	0.99	0.99	0.96

[a]Coefficient of Correlation r of nonlinear correlation between pH and gel time.
Source: Ref. 17.

the higher viscosity and lowered molecular mobility have a stronger effect, reducing reactivity rather than increasing it (Table 5.2). The higher apparent average DP_n value of quebracho tannin might also be due to the small amount of Na metabisulfite (not used in mimosa) used in its water extraction medium, allowing extraction of higher-molecular-weight tannin fractions.

The ^{13}C NMR spectra also give valuable insights into the characteristics of the flavonoid unit B-rings in pine tannins [17]. Thus the bands at 120 to 121 ppm (C6') (Table 5.1) and at 115 to 117 ppm (C5' and C2') (Table 5.1) appear to indicate by both the high values of the sum of their intensities and of the intensity of the C5'–C2' band that quebracho and gambier are much lower in pyrogallol B-rings (Figs. 5.3 and 5.6). By the intensity of the same band (Figs. 5.2 and 5.5), mimosa and pecan nut tannins present a relatively much higher proportion of pyrogallol B-rings [17], a fact also determined by other techniques [19,22]. Pine tannin (Fig. 5.4) appears in this respect to be an intermediate case. This gives an important insight on the reactivity of the tannins. The B-ring reaction with formaldehyde is favored in tannin flavonoid units in which such a ring is pyrogallol. This is because pyrogallol is faster-reacting with formaldehyde than with catechol, as well as for the occurrence of side reactions in units presenting a phloroglucinol A-ring/pyrogallol B-ring (prodelphinidin tannins) combination which renders such a combination particularly reactive [22]. This is the case for pecan nut tannins, as their gel time (Table 5.3) and high strength of unmodified glue lines also appear to confirm [4].

Interpretation of the ^{13}C NMR band intensities for C6' and for C5'–C2' is, however, complicated by the C4–C6 and C4–C8 interflavonoid linkages, in which the C6 and C8 belong to a fisetinidin or robinetidin unit (a unit with resorcinol A-ring). These transmit in the same region. Thus this type of C4–C6 can be calculated to transmit at 121 to 123 ppm, although the band is of low intensity, and this type of C4–C8 is overimposed to the C5'–C2' band and appears to be of more noticeable intensity (Figs. 5.1 to 5.6). However, this confuses the situation mostly in quebracho, where relatively noticeable proportions of C4–C8 interflavonoid linkages coexist with a predominance of resorcinol A-rings.

Two other regions are of interest, the first being the C9 band at approximately 155 ppm (Table 5.1; Figs. 5.1 to 5.6). The higher the intensity of this band, the higher the proportion of flavonoid unit heterocycles maintained close. Thus those (pine and pecan nut) tannins with the highest intensity of this band show very little or no opening of their unit heterocycles because metabisulfite extraction cleaves the interflavonoid link in procyanidin–prodelphinidin tannins preferentially [3,21]. Mimosa and quebracho (Figs. 5.2 and 5.3) show a greater proportion (still small) of cleavage and opening of their unit heterocycles because in profisetinedin–prorobinetinidin tannins this reaction is favored over interflavonoid cleavage and their relative C9 bands are much lower. Gambier tannins (Fig. 5.6), having very low average $\overline{DP}_n$ values, present many fewer interflavonoid linkages to be cleaved; hence the heterocycle is more likely to open, a fact supported by the higher intensity of the C9 band in pure catechin. As in tannin extraction heterocycle opening is not extensive, this band appears to be very sensitive to detection of heterocycle opening and to molecular rearrangements, as are those induced in adhesive intermediates preparation.

A ^{13}C NMR region of interest is the one at 60 to 90 ppm, in which signals occur from carbons of polymeric and monomeric carbohydrates which are always present in industrial tannin extracts. These mask the C2 and C3 bands, which are, instead, clearly discernible in pure catechin (Fig. 5.1). The carbohydrate region is also of importance for an understanding of some of the properties obtained during the preparation of tannin adhesive intermediates [4,17].

In conclusion, the most important differences in characteristics detected—

some already known, others perhaps already perceived, a few not known before—were:

1. Pecan nut tannin extract [4,17]. A higher proportion than expected of both C4–C6 interflavonoid linkages and fisetinidin–robinetidin units. Still predominantly a procyanidin–prodelphinidin tannin but of much more mixed character than pine tannin. The higher proportion of pyrogallol B-rings indicates a relatively high proportion of delphinidin units. It has the highest degree of polymerization of all five tannins examined and little or no heterocycle opening. Proper polymer branching is present.

2. Pine tannin extract [4,17]. Predominantly C4–C8-linked procyanidin–prodelphinidin tannin. Highly polymerized but less so than is pecan nut tannin. A low proportion of both C4–C6 and fisetinidin–robinetidin units. Little or no branching. The relative proportion of pyrogallol and catechol B-rings is intermediate between that of the two tannin classes.

3. Gambier tannin [4,17]. A very low degree of polymerization with a high proportion of monomeric flavonoids. Totally or almost totally catechinic–epicatechinic. A lower proportion of pyrogallol B-rings and a high proportion of catechol B-rings. A higher proportion of open heterocycle form than in the other two procyanidin tannins.

4. Mimosa tannin extract [4,17]. A lower degree of polymerization than those of pine, pecan, and quebracho tannins. A lower number of C4–C8 interflavonoid linkages and a lower number of catechin–epicatechin units than in quebracho. Very predominantly a profisetinidin–prorobinetidin tannin. A higher proportion of pyrogallol B-rings in relation to all the other tannins except pecan nut tannin. A low proportion of branching, lower than that of quebracho tannin.

5. Quebracho tannin extract [4,17]. A higher degree of polymerization than that of mimosa, but lower than that of pecan nut tannin. More C4–C8 and a higher proportion of catechin–epicatechin units than in mimosa. Clearly, a branched tannin: has the highest level of branching other than pecan nut tannin. Lower in pyrogallol B-rings and higher in catechol B-rings. A greater proportion of heterocycle cleavage forms than in pine and pecan nut tannins. Still predominantly a profisetinidin–prorobinetidin tannin.

A similar ^{13}C NMR study [23] describing the modifications introduced by chemical treatment of the structure of tannin during processes traditionally used to produce tannin adhesive intermediates has also been published (see later paragraphs). It confirmed by ^{13}C NMR directly on the natural polymeric tannin aqueous solution several of the findings obtained by model compound studies, but it showed that many of the structural modifications forecast by model compounds occur to an extent too low to be of major significance.

Of particular interest is the system developed by Newman and Porter [18] to characterize by solid-state ^{13}C NMR spectroscopy the semiquantitative proportion of tannins to lignin materials in extracts, and to determine the relative proportions of prodelphinidins to procyanidins in a tannin or tannin extract. They used interrupted decoupling (ID) by pulse sequences to simplify the ^{13}C NMR spectrum to the point at which semiquantitative determinations could be carried out. Two equations [18] were proposed for the determination of the relative proportions of procyanidin, prodelphinidin, and guaiacyl lignin units:

$$\frac{g}{c+d} = \frac{A}{C} - 2.5$$

$$\frac{c}{c+d} = \frac{A}{C} - 2\frac{B}{C} - 0.5$$

where c, d, and g are the concentrations of procyanidin, prodelphinidin, and guaiacyl lignin units, respectively. A, B, and C are the values of the relative areas for the three well-resolved bands across the region of the CP/MAS-ID NMR spectrum assigned to aromatic carbon. Band A was assigned to C5, C7, C9, C3′, and C4′ of procyanidin units; C5, C7, C9, C3′, and C5′ of procyanidin units; and C3 and C4 of guaiacyl lignin units. Band B was assigned to C1′ of procyanidin units, C1′ and C4′ of prodelphinidin units, and C1 of guaiacyl lignin units. Band C is assigned to C4a and substituted C6 and C8 of procyanidin and prodelphinidin units, thus to the interflavonoid linkage.

This system works well, but it does not really indicate what is of greater interest to tannin adhesive technology: (1) the relative proportion of phloroglucinol to resorcinol A-rings in the tannin extract examined (this determines its reactivity toward formaldehyde and hence the type of tannin adhesive technology that should be used), (2) the relative proportion of pyrogallol to catechol B-rings (which is of importance in adhesives cured by tannin autocondensation), and (3) the number-average $\overline{DP}_n$, $\overline{M}_n$, and $\overline{M}_w$ values of the tannin extract under examination.

A much simpler ^{13}C NMR method [24,25] than that of Newman and Porter [18] can be used to determine quantitatively the three types of information needed. In this system ^{13}C NMR analysis is carried out on concentrated (30 to 40%) water solutions of the tannin extract, without an interrupted decoupling pulse sequence—just by a standard ^{13}C NMR spectrum in solution. The relative proportions of repeating units presenting a catechol or pyrogallol B-ring can be determined using the C1′ band at 130 ppm. A single peak indicates catecholic B-rings. If both pyrogallol and catechol are present, the 130-ppm peak splits, with the peak at the higher ppm value corresponding to pyrogallol moieties. The percentage of catechol can then be determined from the relationship

$$\% \text{ catechol B--rings} = \frac{\text{catechol peak}}{\text{(catechol + pyrogallol) peaks}} + c \qquad (5.1)$$

with $c = -0.677$ [24,25].

The number-average degree of polymerization ($D\overline{P}_n$) was found to be determined approximately from the area intensities of the free C8 (95 to 96 ppm) (A), free C6 (96 to 98 ppm) (B), and C4–C8 interflavonoid link (110 to 111 ppm) (C), with the following relationship [24,25]:

$$D\overline{P}_n = 3.708 \frac{c}{A+B}$$

In the same study, the relative average proportion of resorcinolic and phloroglucinolic A-rings was determined for several tannins [24,25]. The ^{13}C NMR bands used were the aromatic C–O for the A-rings (156 to 158 ppm) (A) and for the B-rings (146 to 148 ppm) (B). After determining the A/B peak ratios and expressing this as 1:(B/A), the average number of –OH's on the A-ring will be

$$\frac{\text{average number of}}{\text{--OH's on A--ring}} = \frac{\text{average number of}}{\text{--OH's on B--rings}}{B/A}$$

where the average number of –OH's on the B-rings is determined from equation (5.1). Excellent correlations with experimental results obtained by other means were obtained with this simple equation for the five tannins dealt with in this chapter. The system also worked well for sulfited tannins [24,25].

IV. IMPROVEMENT OF YIELDS IN INDUSTRIAL TANNIN EXTRACTION

Although many different methods of extraction can be employed in the laboratory, the industrial extraction of tannins is generally a fairly simple procedure. The industrial tannin extract is generally a mixture of poly- and monoflavonoids, with a noticeable or even considerable proportion of nonphenolic materials, mainly simple sugars and polymeric carbohydrates. However, not only are different tannins often extracted with slightly different procedures, but some important conditions must be kept in mind to obtain an extract that is at least reasonably suitable for wood adhesives application.

Extraction is generally conducted by placing the comminuted vegetable material in a series of enclosed or covered vats called autoclaves (a misnomer, as pressure is not applied) and extracting with water in a countercurrent so that the extracting solution is considerably enriched as it progresses from vat to vat. In mimosa, only water is used; in quebracho, water containing up to 5 to 10% sodium sulfite or metabisulfite; in pine, water containing 2 to 3% sodium sulfite or metabisulfite; in pecan nut, 5% sodium sulfite with 3 to 4% sodium carbonate or bicarbonate. The initial temperature of the extraction water used also varies: for mimosa, temperatures of 94 to 100°C are used; in quebracho, pine, and pecan nut tannins, temperatures of not more than 70°C are used. To use higher temperatures does not improve the industrial yield of the phenolic material: it often differentially favors extraction of nonphenolic materials. For instance, in mimosa, temperatures of the water of extraction higher than 97 to 98°C (even higher than 100°C reachable by steam back pressure in the autoclaves) improve the yield of total extract quite noticeably but also decrease noticeably the actual percentage of phenolic material usable for adhesives [14,26]. In the other tannins, temperatures higher than 70°C also increase the total yield of extract but also often induce structural modifications in the tannin that might impair, in part, its performance as an adhesive [14]. Thus uncontrolled temperature increases, although giving increased total yield, are not a recommended manner of improving the yield of material usable for wood adhesives. The yield of the four main tannin extracts is, as a consequence, quite fixed in well-determined ranges. Typical mimosa extraction yields are of approximately 30 to 33% by mass of extract (not tannin) of the original bark extracted; quebracho, 26 to 29%; pecan nut, 42 to 43%. The yields of these tannins are comparable and are very acceptable. The yield of pine tannins is, however, somewhat lower than that found using traditional methods of extraction. This is one of the drawbacks regarding their utilization in developing industrial resins.

The best industrial extraction yields in a water medium for pine tannins have been 13 to 15% [27,28], while for mimosa these are 30 to 33% (this means that 300 to 330 kg of tannin extract is obtained for 1 ton of dry mimosa bark). Many routes have been tried to improve this extraction yield, the one presently used industrially giving the yields described above using a mildly sulfited water medium. Although giving higher yields, organic solvent extraction has proven expensive and unacceptable at the simple technological level of an extraction factory. Although some laboratory results have indicated higher water extraction yields, in most cases what was increased was the extraction of lower- and higher-molecular-weight carbohydrates. The amount of tannin did not change and the percentage tannin in the extract, notwithstanding the higher extraction yield, was lower. The reason for the lower extraction yield for pine tannins, even in pine species whose bark is particularly rich in tannins, has been shown to be due to high autocondensed tannin and tannin-derived products in the extract. These are present preceding the extraction process and are often induced in part by the extraction process itself.

Phlobaphenes are reddish water-insoluble phenolic substances which are related to tannins. Phlobaphenes can easily be precipitated from a water solution of polyflavonoid tannis by acid-induced condensation reactions. In the leather tanning industry they constitute the well-known "tanner's red" precipitates which accumulate in leather tanning pits. Some phlobaphenes are formed naturally and are complex mixtures of high-molecular-weight condensed tannins, which have sometimes been found associated with carbohydrates and which contain higher proportions of methoxyl groups [29]. There is speculation as to the differences between the phlobaphenes formed naturally, which could include extraneous matter such as lignans and which are therefore variable, and those formed by acid treatment. Under acid conditions phlobaphene formation is the predominant reaction [6]. In strongly alkaline conditions, partial autocondensation also takes place [6]. As regards their structure, biflavonoids of the (−) robinetinidol/(+)-catechin type and their (−)-fisitinidol homologs, prototypes of mimosa and quebracho tannins, are to a varying degree subject in strong acids or in alkaline solutions (the latter in the presence of $NaHCO_3$–Na_2CO_3 buffers) to positional isomerization, thus giving so-called "phlobatannins" [6]. Their structures, obtained by molecular reorganization, are suggestive of the possible transformations that could be taking place in phlobaphene formation and are shown in Fig. 5.7. The composition of the rearranged molecule led Roux [6] to postulate that these factors contribute to the reduced water solubility of phobaphenes, as has been observed.

Evidence [14,28] suggests that a large proportion of pine tannins remain in the bark even after extraction because when extracted with an organic solvent such as ethanol, the extraction process has been shown to extract effectively much higher quantities of pine tannin [14,28,30–32]. Since pine tannin phlobapehenes dissolve in ethanol, the indication is that during the heated aqueous extraction of the bark, phlobaphenes may form within the

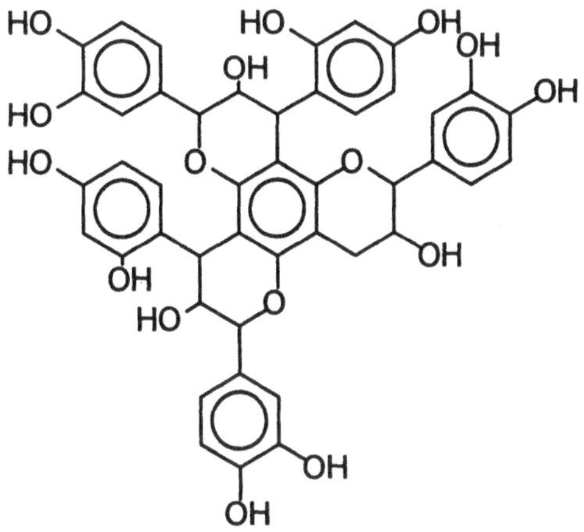

Figure 5.7 Structure of phlobatannin obtained my molecular rearrangement of a flavonoid trimer. (From Ref. 6.)

bark and not be extracted. In addition, air oxidation of the tannin occurs during growth of the tree and after the bark has been stripped from the trunk. Both of these phenomena allow partial self-polymerization of the pine tannin to occur, resulting in increased molecular weight and degree of polymerization and the formation of higher-molecular-weight fractions within the bark. Two approaches can be taken to increase the extraction yield:

1. Minimize autocondensation and self-polymerization of the tannin during extraction in a water medium.
2. Find a solvent to extract all the oxidized and rearranged higher-molec-ular-weight products.

The use of solvents other than water on an industrial scale can be problematic, especially in the areas of pollution, recycling, and high cost. For these reasons, the second approach, which is known to give higher yields [14,28,30–32], is not discussed. The first approach is based, instead, on the apparent characteristic presented by pine tannins to self-condense more rapidly and at lower temperatures than do mimosa tannins [27]. Considering the proposed rearrangement in Fig. 5.7, the presence of the phloroglucinol A-ring, of higher nucleophilicity than the equivalent nuclei in mimosa, might explain the higher sensitivity of pine tannin to autocondensation under milder acidic and milder temperature conditions than those for mimosa tannin. The rearrangement and self-condensation mechanism implies that at some stage during the reaction an attack by the rearrangement intermediates on the nucleophilic A-ring of a neighboring flavan unit by promotion of a positive charge or by another mechanism does occur (Fig. 5.7). Blockage of this attack by the introduction of a more reactive species could then lead to stoppage of the self-condensation reaction and avoidance of the formation of phloba-phenes. As self—condensation implies condensation with the highly reactive phloroglucinolic A-rings of pine tannin flavonoids, reactive species that can be used to stop such self-condensation, must be equal or superior in nucleophilic action to the phloroglucinol A-ring of the pine tannin. In this case, phloroglucinol, m-phenylenediamine, and urea were used [14]: the former two exclusively to prove that the mechanism of self-condensation could be blocked, the latter both for the same purpose and because its low cost would render it of interest at the industrial level. This approach would be apt to stop self-condensation of the tannin only during and after extraction, decreasing molecular weight and viscosity and increasing the extraction yield of the tannin.

A. Mechanism

Many mechanisms of tannin self-condensation and rearrangements and of phlobaphene formation have been proposed. If in the case of the rearrangement of pine tannins, the mechanism proposed [6] might be valid, the probable mechanism of stoppage presented here can be schematized as shown in Fig. 5.8. The proportion of urea present will determine if the self-condensation to phlobatannins is stopped or still proceeds by reaction of other cleaved flavonoid units on the remaining available reactive sites of urea. Equally, a fast-reacting phenol such as phloroglucinol could also function as a blocking agent of the self-condensation and rearrangement reaction. m-Phenylendi-amine should react as both urea and phloroglucinol. If, instead, the mechanism that is valid is the one proposed by Freudenberg and de Lama [33] for acid opening of the flavonoid heterocyclic ring, blockage can be as represented in Fig. 5.9. This reaction competes with cleavage of the heterocyclic ring by sulfitation in tannin extraction.

If, instead, the mechanism proposed by Hemingway et al. [21] for interflavonoid bond cleavage, particularly as applicable to pine tannins, is valid, addition of a strong nucleophile should block the self-condensation reaction. Hemingway et al. [3,34] determined that the interflavonoid bond of procyanidins is extremely labile to cleavage. They suggest that significant interflavonoid bond cleavage would be expected at ambient temperature and pH 5.0 [3,34]. The quinone methide produced will again react with a

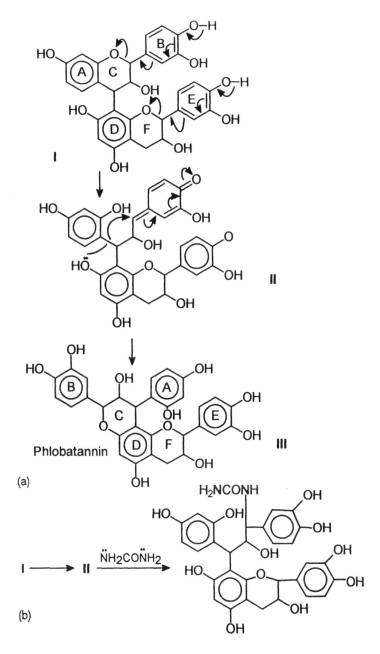

Phlobatannin

(a)

(b)

I ⟶ II

Figure 5.8 (a) Mechanism of rearrangement of a mimosa flavonoid dimer to phlobatannin (from Ref. 6); (b) suggested mechanism of urea action in blocking the rearrangement to phlobatannin.

173

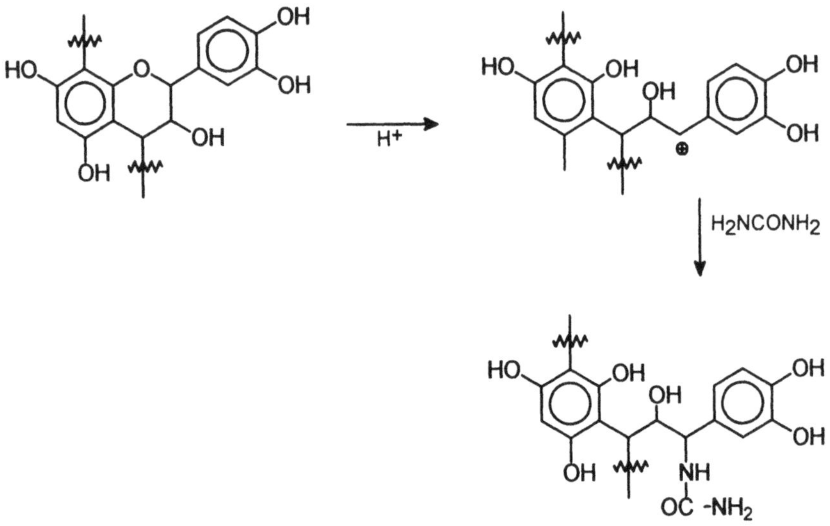

Figure 5.9   Mechanism of cleavage of flavonoids etherocyclic ring by acid [33] and of the suggested urea reaction with the flavonoid carbocation formed to block tannin autocondensation.

nucleophile, such as the phloroglucinol A-ring of the same or a neighboring flavan unit [3]. In the case of the probable mechanism of stoppage of self-condensation presented, if a strong nucleophile or a more mobile strong nucleophile is added, the quinone methide will react preferentially with it, blocking the self-condensation reaction of the tannin. The extent of blockage will depend on the amount of strong nucleophile added. Blockage by addition of resorcinol has already been demonstrated [3,21,35,36]. Addition of phloroglucinol and m-phenylenediamine should give similar products. Inhibition of self-condensation by urea might proceed as shown in Fig. 5.10. All these proposed self-condensation mechanisms for tannins have been presented, whatever their relative merits and likelihood of existence under the relevant reaction conditions, to show that the addition of a strong nucleophile will, whichever of these mechanisms is occurring, block self-condensation of the tannin to phlobaphenes partially or completely during and after pine tannin extraction.

A recent study of industrial-scale *Pinus radiata* tannin extraction has shown that this is the best approach to use for pine tannins, at least for *Pinus radiata* tannins [14]. The system used water extraction to which sodium metabisulfite together with urea were added. The yield improved from the 13 to 15% typical

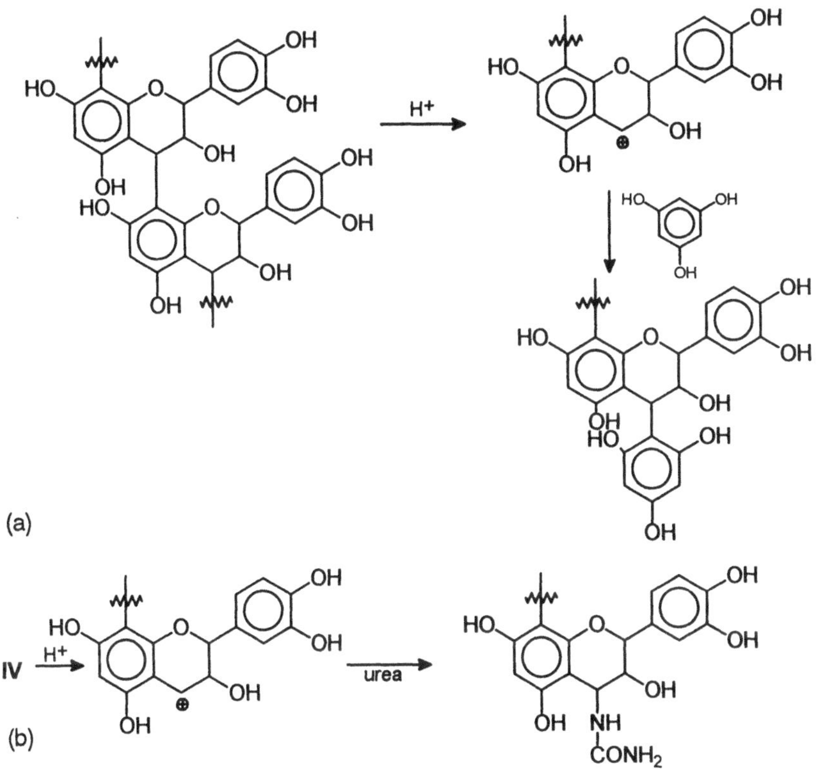

(a)

(b)

IV

Figure 5.10 (a) Mechanism of cleavage of pine tannin interflavonoid linkage [3,21] in acid and of the blocking of the reaction of tannin autocondensation of phloroglucinol or resorcinol; (b) suggested mechanism of urea action in blocking the carbocation formed by the cleavage of interflavonoid linkages and of its blocking of tannin autocondensation.

for water + metabisulfite extraction, to 18 to 20% [14,23], which is a definite improvement as to the cost of pine tannins.

The results of three layers of laboratory particleboard made using pine tannins in which phlobaphene precipitation was stopped by urea addition in the water–sulfite medium were also checked. The results obtained indicated that with the formulations used, the results obtained are comparable to those of standard exterior-grade tannin boards, pine or mimosa tannin-bonded; the presence of small amounts of urea did not appear to influence the characteristics of the board. Internal bond (IB) strengths are rather marginally higher than those obtained with non-urea-containing tannin extracts.

V. FORMULATION OF NONFORTIFIED TANNIN ADHESIVES

Tannin–formaldehyde (TF) adhesives that are not fortified by the use of a synthetic resin are the most commonly used adhesives in exterior particleboard application [1,4,23,37]: some applications for plywood are also on record [1]. The older nonfortified thermosetting TF adhesives for particleboard rely on treatment with chemicals to achieve structural modification of the tannin to increase its reactivity. In reality, such modifications were introduced in the late 1960s and early 1970s primarily to reduce the viscosity of the tannin extract to a level capable of being handled by the factory glue blenders of that time. Improvement in the performance of the adhesive by such treatments was, although still considered significant, of less importance than the ability of the process to decrease the viscosity of the tannin solutions [1,56].

Nonfortified tannin adhesives formulations used today can be divided into two main classes: formulations based on chemical modification of the tannin extract preceding its use in the TF mix, and formulations based on the nonmodified tannin extract. The latter were introduced with the advent of the faster-reacting procyanidin–prodelphinin tannins such as pine and pecan tannin. These formulations are discussed first, not only because of their interest, but because they will help to put into perspective the significance, or lack of it, of the earlier, more established and extensively used modified tannin formulations.

Nonfortified, nonmodified tannin extracts, rather than chemically modified tannin adhesive intermediates, can be used effectively to produce excellent exterior-grade particleboard at fast pressing times of industrial significance, with considerable advantages in both handling and cost. This can be achieved by simple pH-controlled reactivity adjustments of the tannin extract in the glue mix. This concept is shown to be applicable to all types of polyflavanoid tannin extracts: pine, mimosa, pecan, quebracho, and gambier tannin extracts are compared as to performance and capability. Applied to the faster-reacting tannins, such as pine tannin and pecan tannin extracts, the concept is shown to produce fast hot-pressing rates and excellent high-moisture-content tolerances which are undreamed of for the slower-reacting tannins, modified or nonmodified, traditionally associated with thermosetting wood adhesives. Techniques used industrially to control well viscosity and pot life of the glue mixes of accelerated tannin adhesives are also presented.

A. Nonmodified TF Adhesives for Particleboard

Tannin-based adhesives for exterior-grade particleboard based on mimosa tannin extract and to a lesser extent on quebracho tannin extract have been in commercial operation in several southern hemisphere countries since 1971 [38]. The adhesives based on these two tannins are much slower reacting with formaldehyde than are the newer generation of adhesives based on the faster-reacting procyanidin–prodelphinidin tannins such as pine bark and pecan nut pith tannin extracts introduced since 1990 [4,38]. The structural reasons for the differences in reactivity toward formaldehyde of the two slow-reacting tannins, mimosa and quebracho, and two faster-reacting tannins, pine and pecan, are well known. They are based on structure of the A-rings of the majority of flavonoid units composing the two sets of tannins: resorcinol-like in mimosa and quebracho, phloroglucinol-like in pine and pecan nut tannins [1,4]. Their reactivity has always been thought to put the fast tannins at a disadvantage in relation to the slower-reacting tannins used traditionally for wood adhesives [1] because hardening reactions under industrial conditions were thought not to be easily controllable, leading to problems of precuring, particularly because of the traditional operating pH of tannin adhesives, causing the pot life of the adhesive glue mix to be too short [1]. Furthermore, because the primary tannin adhesive based on mimosa was based on structural modifications of the tannin extract itself [37], without synthetic resin fortification, the ineffectiveness on the fast tannins of such modifications appeared to deny their use as lower-cost nonfortified exterior-grade adhesives. Instead, the high reactivity of these tannins permits application feats in wood adhesives which are impossible to achieve with the slower, traditional tannins. They also show that the misconception that any flavonoid tannin, slower or faster reacting, needs to be chemically or structurally modified, or fortified by addition of synthetic resins, to perform as an excellent thermosetting wood adhesive needs to be corrected. It puts the success of an exterior tannin adhesive for particleboard firmly in the field of its conditions of application rather than in that of its chemical modification.

In many applications of the faster-reacting tannins [4,39,40], indications are that the reactivity (hence the rate of the reaction of the tannin with formaldehyde), is the only performance-determining factor for tannins used as thermosetting wood adhesives. This implies that for all polyflavanoid tannins it is of no or little final importance if modifications to the tannin structure are [4,17,23], or are not, made. What is important is that the correct reactivity, hence the correct cure rate, be obtained. This is easier to achieve simply by varying the pH of a nonmodified extract rather than going to the expense and bother of modifying the tannin extract chemically to form an "adhesive intermediate" [37,40]. At pH values of equal reactivity there is hardly any difference in the internal bond (IB) strength results, particularly in IB strength after a 2-h boil conducted according to the DIN68763 (V100) test (Figs. 5.11 and 5.12) [41]. Thus there is no difference between particleboard made with a modified (more expensive) tannin adhesive and one made with plain, nonmodified tannin extract if its reactivity toward formaldehyde has been adjusted (by pH) to be the same. The only two exceptions to this trend are pecan nut tannin extract, which presents better results, an occurrence depending on the presence of additional cross-linking autocondensation reactions already described [4]; and quebracho extract, which is known to be slow reacting.

Figures 5.11 and 5.12 compare the gel times with formaldehyde of five tannin extracts usable for adhesives: from these figures the pH values of comparable reactivities of the five tannins, at which to obtain similar particleboard results, can easily be derived. A few important conclusions can then be drawn:

1. Reactivity toward formaldehyde, hence gel time, is all that counts in preparing a tannin adhesive of excellent performance under a given set of application conditions.

2. It is useful but in reality unneccessary to modify the tannin extract chemically as has now been done for decades [1,37] to prepare, for instance, a mimosa tannin adhesive intermediate. Consequently, modifications determined exclusively by model compounds, such as the phlobatannin rearrangement, which have mistakenly been believed to be potentially important in adhesive application [6], become, instead, rather inconsequential to the field of thermosetting tannin adhesives for use with wood. In most cases it appears that equally good particleboards can be produced simply by adjusting the pH of the nonmodified tannin extract in the glue mix at ambient temperature to the reactivity desired [40]. This is valid for all five tannin extracts investigated.

3. There is no doubt that in the quest for faster pressing times and improved tolerance to higher moisture content of the resinated wood furnish, the fast-reacting tannins, pine and pecan, have more advantages and greater flexibility over the slower-reacting ones such as mimosa and quebracho. From Figs. 5.11 and 5.12, even pushing reactivity to its limit, the structurally modified (hence "faster") mimosa and quebracho adhesives can never hope to reach gel times as short and reactivity as rapid as those achievable with pine and pecan nut tannin extracts. This becomes even more evident (Fig. 5.12) when the slower, nonmodified tannin extracts are compared to the faster ones. The curves in Fig. 5.12 indicate that the results obtainable with pine at

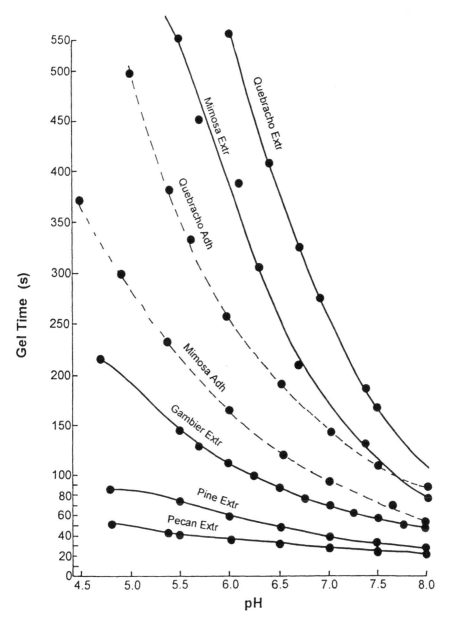

Figure 5.11 Gel time as a function of pH of five nonmodified tannin extracts and two modified tannin adhesive intermediates (dashed lines) in the pH range 4.5 to 8.0.

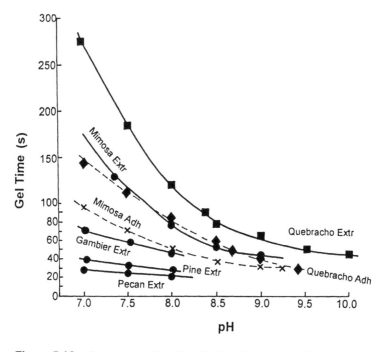

Figure 5.12   Gel time as a function of pH of five nonmodified tannin extracts and two modified tannin adhesive intermediates (dashed lines) in the pH range 7.0 to 10.0.

pH 7.7 or higher cannot easily be matched by mimosa or quebracho even at pH values over 8.5 and 10, respectively, where their gel-time curves finally flatten out.

It is clear from the above that only three conditions can then limit industrial application of such a new concept: (1) the pot life of the glue mix, (2) the viscosity of the glue mix, and (3) the rate of moisture elimination from the board in the form of vapor during pressing. The first point is easily addressed: it is the combination of tannin extract and formaldehyde at the pH needed that is the cause of the problem. It is then sufficient to subtract just one of these elements from the glue mix and add it separately to solve such a problem with ease. This is already done in several of the factories in which tannin adhesives have been used or tried. In some factories the problem is solved by adding separately fine paraformaldehyde powder, by means of a small screw conveyor, to the wood particles before the glue blender in which the liquid tannin extract solution at high pH is blended to the furnish [42,43]. In other factories, the liquid glue mix is composed of a solution of the tannin extract at its minimum pH value of reactivity (4 to 4.5, generally the tannin natural pH) mixed with paraformaldehyde powder hardeners; this mix has a potlife in excess of 30 h. A measured stream of diluted sodium hydroxide is also added: directly to the glue blender through a separate line [44] or by means of a short static mixer to the line carrying the liquid glue mix, just a few centimeters before the glue blender, to adjust reactivity. All these are simple industrial practices, the former two having been in daily industrial use now for several years.

The viscosity of the glue mix also does not constitute a problem. In addition to factory techniques already reported and used for a long time [43], several others are now in use. The chemical modification of tannin extracts was originally prompted by the need to have lower viscosity at the highest solid content possible [1,37]. The introduction of the accelerated reaction of tannins induced by the higher pH's faster development of bond strength at an earlier stage of the cycle during hot pressing of the board ensures that the high-pressure steam generated within the board at higher moisture contents is not capable of breaking the strong bonds already formed. In short, disruption by steam emission of bonds while they are being formed is minimized. Thus excess moisture content no longer constitutes a problem; viscosity can easily be adjusted by adding water, dropping the solids content of the adhesive mix [40] (as low as 30% for pecan and as low as 35% for pine tannin extract). This is possible because these bonding systems have now become very tolerant of very high moisture contents. Another system already used industrially [39,40] involves warming the liquid glue mix up to 30 to 35° C: its viscosity then decreases precipitously. Pot life is not affected because the liquid glue mix does not contain the paraformaldehyde hardener, or if it does, the tannin is already at its pH of minimum reactivity.

The only limiting factor in how fast a pressing time can be achieved, then, is the rate of elimination of moisture from the board during hot pressing; after all, no one wants too high a moisture content in the finished panel, as this would cause dimensional instability, warping, and other problems. A particleboard manufacturing process using a variant of the high-pH system described above maintains the moisture content of the resinated particles above 20% at standard press times of 9 s mm^{-1} (190° C). Press times as fast as 7 s mm^{-1} have been achieved [42] even with one of the slower-reactivity tannins. It is clear, then, that by the use of higher pH values and of the fast-curing tannins, it is quite likely that industrial press times faster than these are achievable. It is also quite clear that the slower-reacting tannins could be somewhat upgraded by the addition of percentages of the faster-reacting tannins, a fact evident from the curves in Figs. 5.11 and 5.12.

The excellent particleboard results obtained with other fast-reacting tannins then denies the mistaken assumption that at alkaline pH values, catechinic acid rearrangements occur such as to impair their performance as adhesives [9]. It has now been clearly shown that the fast tannins are capable of excellent adhesive performance not only at mildly acid and neutral pH values, but also at alkaline pH values [7].

B. Modified TF Adhesives for Particleboard

Mimosa tannin extract composed of 76% polymeric flavonoids, 4 to 5% monomeric flavonoids, 3 to 6% hydrocolloid polymeric carbohydrate gums, and 3 to 5% simple sugars is normally treated [37] in 50 to 60% water solution with acetic or maleic anhydride, phenyl acetate, and caustic soda to obtain particular effects. Some of the causes of these effects have long been known [1,56], while others have been discovered and reported much more recently [4,23]. In summary, these are:

1. Hydrolysis of the hydrocolloid gums by NaOH and anhydride (which goes to acid) treatment leading to lower extract viscosity [1,2,4] and to a very small number of rearrangements of the phlobatannin type. Hydrolysis of the gums to simple sugars, thus decrease in the proportion of polymeric gums in the extract, improves the strength of the adhesive considerably. The strong effect of polymeric gums even on synthetic PF resins is well documented [1,2].

2. The α-set [4,45,46] acceleration effect on the TF reaction under alkaline conditions obtained by acetylation (or other esterification) is obtained by anhydride treatment of the tannin extract (not of the tannin alone). This effect can only be obtained where consistent amounts of hydrocolloid gums are present in the extract. Such gums have strong colloidal characteristics, forming in water, micelles containing part of the tannin. This is why treatment of 50% water solution of a tannin extract with an anhydride gives partial esterification of the tannin followed by rapid α-set rearrangements in water solution, as part of the anhydride migrates within the micelles where water is not present and tannin esterification occurs [23]. Thus colloidal behavior is essential in the initial part of the extract treatment.

3. The α-set acceleration effect caused by the external addition of an ester as phenylacetate. Both this and point 2 above lead to faster curing and a higher cross-link density in the tannin.

4. The well-known rearrangement caused by cleavage of the inter-

flavonoid 4,8-linkage by both acid and alkaline treatment [21] characteristic of procyanidin tannins and by cleavage of the flavonoid units' pyran heterocycle in tannins such as mimosa. ^{13}C NMR investigations of what occurs in natural polymeric tannin solutions as a consequence of hot modification with the chemicals mentioned above has indicated that a variety of structural rearrangements occur. Some of these rearrangements contribute to enhanced performance of the tannin, whereas others do not.

The primary structural modifications that appear to affect the performance of the modified tannin extracts for adhesives during their reaction with formaldehyde are summarized in Figs. 5.13 and 5.15. Some of these modifications have already been forecast by model compound work, or can be deduced or inferred by applied adhesive work already reported, but were checked only recently for concentrated solutions of the tannins as used in adhesives [4,23]. Model compound findings have also not been related to the extent to which some of these modifications might or might not occur in the tannin itself and to the relative importance to the final performance of a tannin as an adhesive. The importance, or for some tannins generally the lack of it, of some reactions previously believed to be performance determining has only recently [23] been put into its correct perspective directly on the tannin itself and for different tannins. In mimosa, the phlobatannin rearrangement introduces a more reactive resorcinol ring in a configuration of higher mobility. This will lead to faster reactions, as observed by faster gel times (Figs. 5.13 and 5.15) and only in some cases possibly a somewhat higher density of cross-linking due to the resorcinol A-ring being in a configuration of higher mobility. Quebracho behaves very similarly to mimosa, possibly with a more extensive proportion of phlobatannin rearrangement, but also with some depolymerization. The sum of the two effects gives an overall effect similar to, but slower than, that for mimosa (Figs. 5.11 and 5.12) as regards reactivity, indicating that quebracho depolymerization is possibly more extensive than the phlobatannin rearrangement, a fact also indicated by the most noticeable decrease in viscosity.

Their already much higher reactivity being due to the predominant amount of phloroglucinol A-ring in their constituent units, the two more reactive tannins, pine and pecan nut, behave very differently. In the case of pine, the chemical treatments do not really improve the reactivity or degree of cross-linking: the improvements noted are due to reactions in the colloidal state [4,23]. In pecan nut tannin autocondensation (No. 3 in Fig. 5.14) leads to an unchecked increase in viscosity which renders the tannin unusable for adhesives if chemically and heat treated. The fact that the same autocondensation occurs during hot bonding while the tannin–formaldehyde

a) Mimosa (a) small amount of autocondensation at C8

(b) little or no change of interflavonoid link

(c) phlobatannin rearrangement (acid and alkaline induced) main reaction

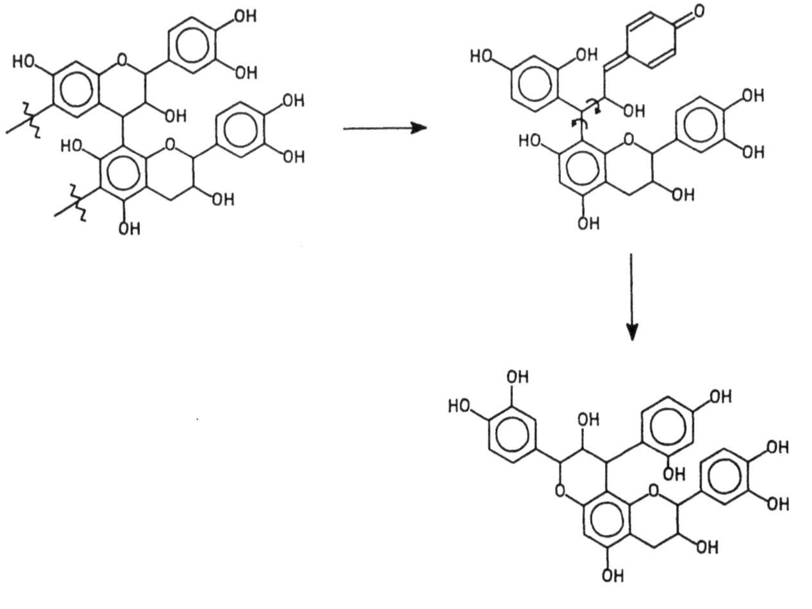

Phlobatannin

(b) Quebracho (a) extensive interflavonoid clevage

(b) depolymerization by C4–C8 clevage at phloroglucinol units

(c) closure of heterocycles by phlobatannin rearrangement

1)

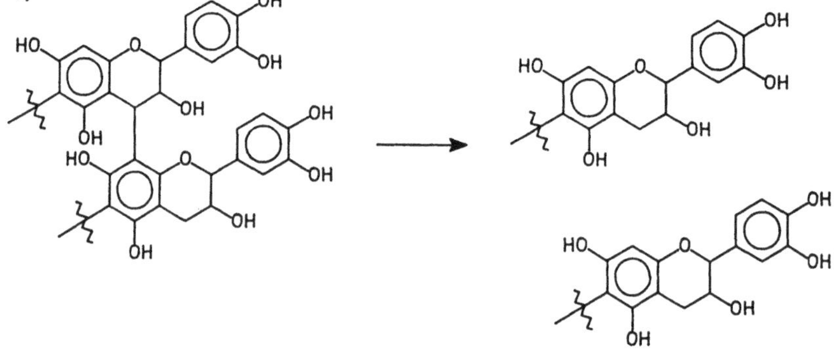

decrease in DP

Figure 5.13 Schematic of main reactions occurring in polymer modification by chemical and heat treatment of mimosa and quebracho tannin to be used for polycondensate adhesives.

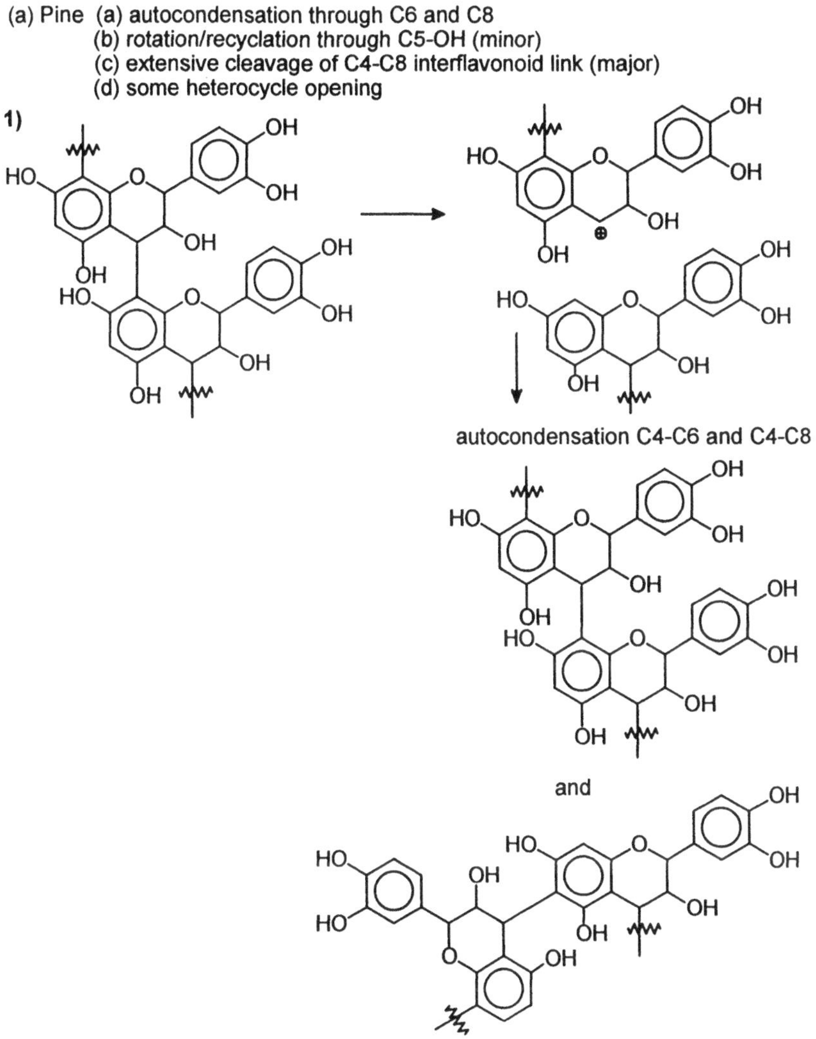

Figure 5.14 Schematic of main reactions occurring in polymer modification by chemical and heat treatment of pine and pecan nut tannin to be used for polycondensate adhesives.

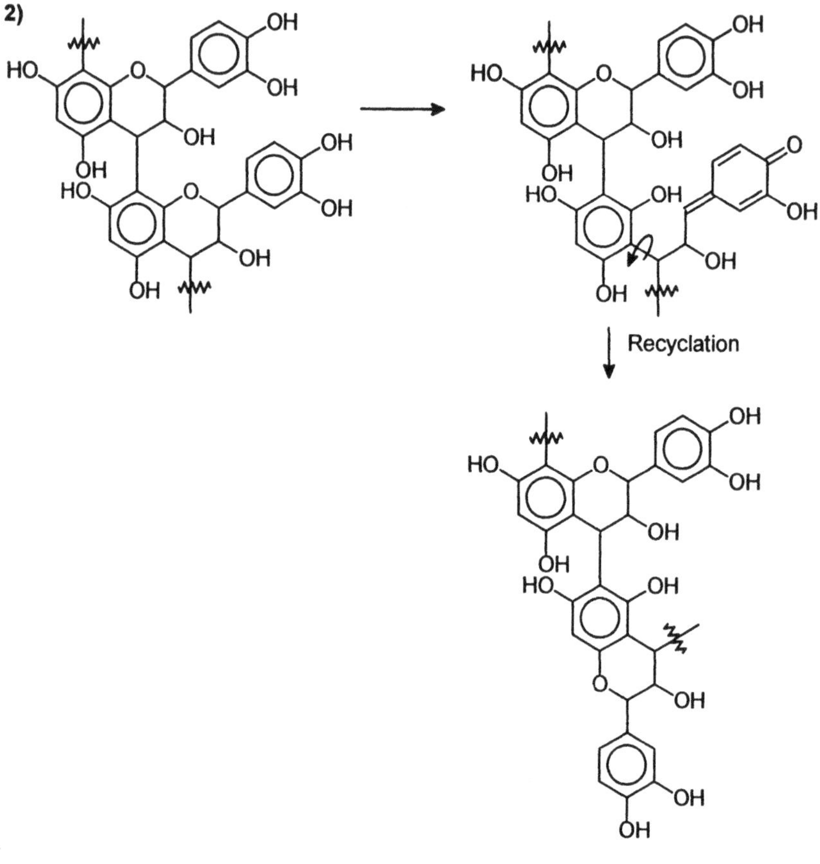

2)

Figure 5.14   (continued)

reaction is ongoing, however, helps this tannin, when used as an adhesive, to outperform all the other tannins in their modified and unmodified forms [4,23,40]. For pecan nut tannins simple treatment with urea [4,14,23] at ambient or higher temperature makes it possible to check the increase in viscosity to a considerable extent.

From the mechanisms at play observed by ^{13}C NMR, pine, mimosa, and quebracho could probably be modified to a certain extent to upgrade their performance to the level of pecan nut tannin. To this effect, addition of some pine tannin or gambier tannin, both catechinic, to mimosa or quebracho will probably upgrade their performance by the same mechanism as that observed

(b) Pecan Nut : as in 1) and 2) in pine and
3)

Phlobatannins as mimosa
(minimal or not occuring)

in pecan nut tannin alone. It will probably work best in upgrading pine tannin performance as an adhesive by the addition of small amounts of mimosa or quebracho extracts to induce pecan nut-like autocondensation on TF adhesive curing.

It is essential to point out the important characteristics of pine bark tannin, pecan nut tannin, and mimosa tannin extracts, as they are available for commercial use (Table 5.4). The single most important characteristic in which pine extract and pecan nut extract differ is the content of hydrocolloid gums. The characteristics of the three tannin extracts shown in Table 5.4 indicate

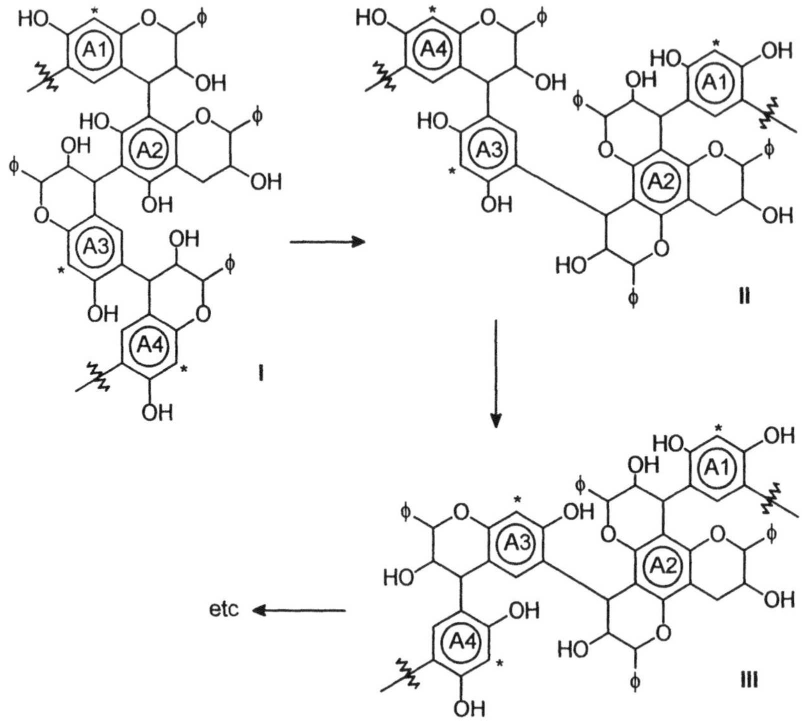

* sites highly reactive with formaldehyde. Number remains the same throughout
sequential rearrangement steps
φ = Flavonoid B-ring

Figure 5.15   Schematic of limit of improvement caused by phlobatannin rearrangement in tannins to be used for polycondensation adhesives.

that the viscosity component due to hydrocolloid gums is likely to be high in pine bark and mimosa bark extracts and low in pecan nut extract. In the latter the high viscosity appears to be due primarily to the high-molecular-weight tannins. Thus in mimosa extract the viscosity appears to be due mostly to hydrocolloid gums and much less due to tannin. In pine extract the viscosity contributions of gums and tannins appear to be roughly comparable. In the pecan nut extract the viscosity appears instead to be due almost exclusively to the high-molecular-weight tannins.

By applying to these three different tannins the standard chemical treatment discussed, it would be logical to expect (1) a decrease in viscosity for mimosa

Table 5.4 Typical Values for Three Types of Commercial Tannin Extracts

	Pine bark tannin extract	Pecan nut tannin extract	Mimosa bark tannin extract
Phenolic content (%)	±80	±87	±80
Approximate number-average degree of polymerization	6–8	8–10	4–5
pH	±4.8	±5.6	±4.2
Hydrocolloid carbohydrate gum content (%)	5–8	0.0	3–6
Simple sugar content (%)	3–5	6.9	3–5
Typical extract viscosity, 40% solid content, at 25°C (CP)	230	520	150
Extent of sulfitation in extraction (%)	2.5	2.5–5.0	0.0
Gel time at 100°C and pH5.3 (s)	55	40	557

Source: Ref. 4.

bark extract, (2) a decrease in viscosity for pine bark extract, and (3) no noticeable variation in viscosity for pecan nut extract. What is observed instead is a decrease in mimosa extract viscosity, no great variation (but a decrease) of pine extract viscosity, and a considerable increase in the viscosity of pecan nut extract [4]. Even more interesting is the effect of the standard treatment on the gel time at the standard pH values of operation of the three tannin adhesives. The treatment leads to faster gel times for mimosa and pine but does not appear to have any effect on the reactivity of the pecan nut extract. Furthermore, the results of internal bond (IB) strengths of particleboard obtained by using the three types of tannin adhesives also appear to follow an interesting trend: that the standard treatment improves the performance of the adhesives derived from pine (slightly) and mimosa extracts but does not really improve the performance of adhesive prepared from pecan nut extract. In pine bark tannins and mimosa bark tannin acid, alkaline hydrolysis of the hydrocolloid gums during treatment leads to the characteristic increase in strength. The proportion of hydrocolloid gums on any phenolic adhesive, including tannins, has a marked effect on both the original strength and the water resistance of the adhesive. Curves of the effect of colloidal gums and simple sugars have been reported, and the effect can be summarized as shown in Fig. 5.16.

Hydrolysis of the gums is then likely to shift the relative balance of phenolic materials, gums, and simple sugars such as to improve the cured strength of the adhesive. This improvement in the cured strength of the adhesive cannot occur for pecan nut extracts because it contains hardly any gums and a low

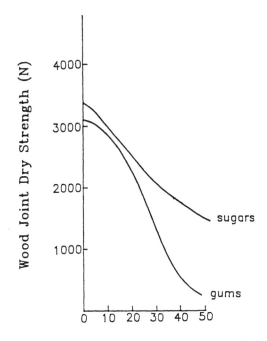

Proportion of sugars and gums (%)

Figure 5.16 Comparative effect of simple sugars and hydrocolloid gums on the strength of a cured PF resin.

percentage of sugars. Equally, acetylation of the tannin, due in pine and mimosa to the colloidal state imparted to the solution by the gums, cannot occur in pecan nut tannins. The lack of acetylation and subsequent α-set, the already high reactivity toward formaldehyde of the pecan nut extract, and particularly the range of pH values at which it can be used in its adhesive applications forbid the existence of an α-set cure acceleration mechanism for this extract. Thus, after treatment, both gel time and cured adhesive strength values remain practically unaltered.

On the basis of all the above, one would expect no variation of viscosity before and after treatment of the pecan nut tannin extract. However, the interflavonoid bond cleavage [13] and particularly, heterocycle cleavage [4,22,23] which prodelphinidin-type tannins undergo in an acid or alkaline environment, thus present in both pine bark and pecan nut tannins, causes cleavage of polyflavonoids. For instance, for interflavonoid linkage cleavage:

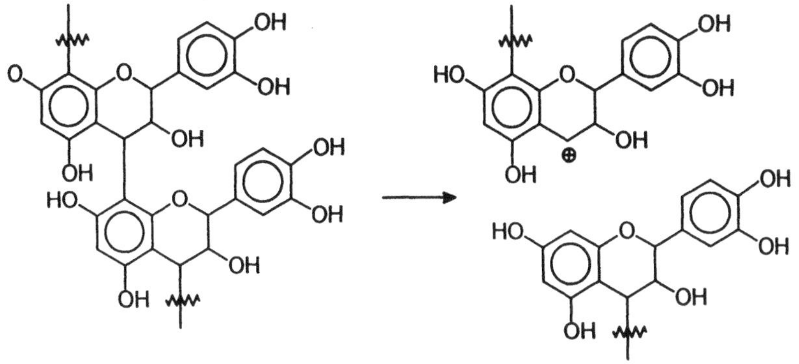

It also leads to autocondensation through the carbocations formed:

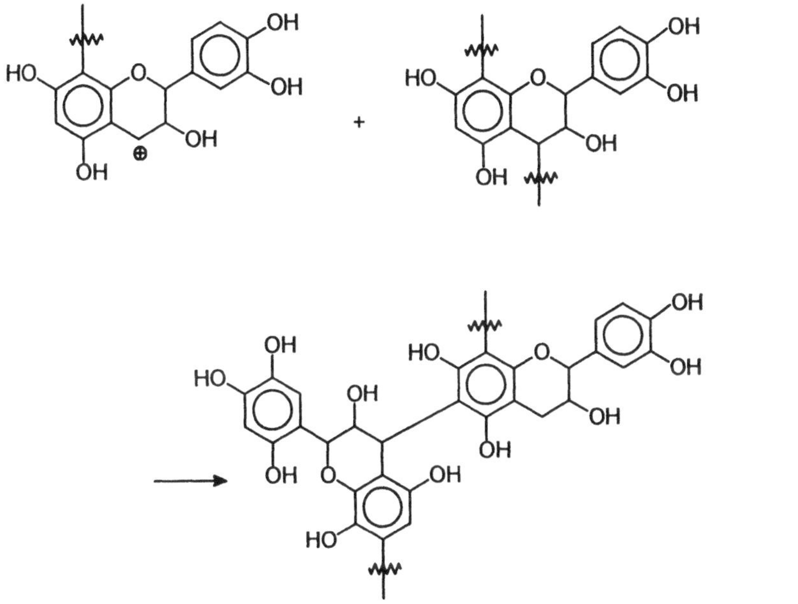

Mono- and polyflavonoid fragments have a low probability of recombining to polyflavonoids having the same degree of polymerization as the starting material. The branching coefficient of a polymer is the probability that a given branched unit will join another branched unit rather than an end group. At values of the branching coefficient equal to 0.5 or lower (according to monomer functionality), the molecule is a continuous chain theoretically equivalent to a gel [47]. In reality it is likely that fewer chains of higher degree of polymerization (DP) than the starting ones are formed, this translating to a much higher degree of entanglement and hence higher viscosity. This hypothesis is also supported by the increase in viscosity, standing at ambient temperature, of 40% solutions of pecan nut tannin extract. The process is accelerated by heat and by the standard acid + alkaline modification used to prepare a tannin adhesive.

Another important question is: Is what is observed a true increase in the $\overline{DP}_n$, or is it the formation of inherently labile colloidal structures, as observed, for instance, in the phenomenon of thixotropicity? Part of the phenomenon is also due to this factor. Thus for pecan tannin extract (1) the permanent increase in viscosity due to autocondensation accounts for only about half of the total observed increase in viscosity [4]; (2) the remaining component of the increase in viscosity is due to the longer autocondensed tannin molecules interassociative forces, which have increased due to higher degree of entanglement [4]; (3) the use of urea is an effective way to block autocondensation [4]; and (4) this extract is perfectly usable industrially [4]. It must be pointed out that not all pecan nut tannin extracts present such unusual behavior. Most of them maintain a very stable viscosity, but due to the high molecular weight of the tannins, such behavior can sometimes be observed.

In the case of pecan nut tannin extract, considering that due to lack of a colloidal state, acid–base treatment should not greatly benefit the tannin adhesive, such tannin should then be able to function well as a tannin–formaldehyde adhesive for particleboard without chemical modification of the extract. Results obtained support this view, and the IB-strength results of particleboard prepared with a nonmodified pecan nut extract are all good [4,40]. In the case of the pine tannin extract, two competing trends are present: (1) the decrease in viscosity caused by gum hydrolysis, and (2) the cleavage and recombination of the tannin fragments leading to higher viscosity. It appears that the former is more marked and hence the viscosity decreases somewhat, but not too extensively.

VI. FORMULATION OF TANNIN ADHESIVES FORTIFIED WITH SYNTHETIC RESINS

A. Fortification with Formaldehyde-Based Resins

The oldest type of tannin adhesives used for thermosetting applications use the addition of small amounts of synthetic resin to the tannin. This approach was favored at the beginning of tannin adhesive research because first, adhesive chemists were more comfortable with it, as the technology involved presented at least some similarities with the synthetic resins they were used to. Second, it counterbalanced the deleterious influence of the carbohydrates present in the tannin extract. There is no doubt that this approach works, but adhesives prepared in this manner cost more and often present grave and unwanted technical drawbacks. Thermosetting tannin adhesives fortified with synthetic PF resols, or even with a few percent resorcinol, or with resorcinol–formaldehyde or phenol–resorcinol–formaldehyde resins, are now completely out of use, nor did they ever gain a significant industrial market. These systems have already been described extensively elsewhere [1,37]. Much more successful and still used extensively industrially are the exterior-grade tannin adhesives obtained by fortification of the extract with aminoplastic resins, in particular with UF resins. These formulations are still used extensively for exterior-grade plywood [37,48], as waterproofers of starch adhesives for corrugated cardboards [49], and for tannin-based acid-setting foams [50]. They never gained any industrial foothold in particleboard, although they also work well for such applications. Although these systems have been described extensively elsewhere [1,37], it is of interest to discuss in more depth the rationale behind the copolymerization of UF resins with tannins. Although mimosa tannin will be taken as an example, similar considerations can be applied to the other polyflavonoid tannins once due consideration is given to their lower or much higher reactivity toward formaldehyde and methylol groups.

The exact ratio between copolymer and polymer blend produced with these adhesive systems depends on such factors as the proportion of the individual reactant monomers used and their relative reactivities [50]. In polycondensates the relative reactivities of the constituent species can be related directly to their gel times, and therefore an approximation of the ratio of copolymer to polymer blend can be predicted from gel times versus pH curves for two-component polycondensate, thermosetting resin systems [50].

In the case of tannin-based adhesives, as well as for other applications, consider a system consisting of both a mimosa TF resin and a UF resin. The gel times of each resin by itself as a function of pH are shown in Fig. 5.17. Equally, the rate constants of the self-condensation of each resin, also as a function of pH, are shown in Fig. 5.18. Considering, first, the UF resin curve (Fig. 5.17) it can be seen that the gel time increases exponentially, and therefore the rate of self-polymerization decreases exponentially (Fig. 5.18), as the pH moves through 1 to 7. The TF curve indicates that a minimum rate of condensation is experienced at a pH of 3.4, with an increase in condensation

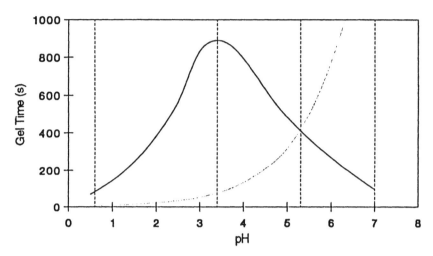

Figure 5.17   Gel time of mimosa TF (solid curve) and UF resin (dashed curve) as a function of pH.

rate as the pH becomes more acid or alkaline. Finally, in Fig. 5.18 it can be seen that the formation of methylol urea from any free urea and formaldehyde (U + F → UF) has a minimum rate of reaction in the pH range 5 to 8.

Comparing the two curves in Fig. 5.17, it can be seen that in the alkaline region (above pH 7) the rate of self-polymerization of the UF resin is negligible compared to the rate of mimosa tannin reacting with formaldehyde to form a methylol group which then reacts with another tannin nucleus. The methylol condensation is due to the alkali activation of the tannin nuclei, and therefore any urea methylol groups will react preferentially with the phenolic rings of the tannin to produce $-NH-CH_2-$ tannin bridges. The implications of this are that a mimosa TF–UF resin system in an alkaline environment (i.e., above a pH of 7) will produce a hardened product that will consist primarily of copolymers with a negligible amount of polymer blend [50].

As the pH decreases, the reaction rate of self-polymerization of the UF resin increases to pH 5.3, where it intersects the reaction rate of the mimosa tannin–formaldehyde resin which has been decreasing throughout this pH range (Figs. 5.17 and 5.18). The result of this decrease in pH is an increase in the percentage of polymer blend, with a corresponding decrease in the percentage of copolymers. This can be seen as indicating that the proportion of $-NH-CH_2-$ tannin bridges decreases with a corresponding relative increase in the proportion of $-NH-CH_2-$ NH– and therefore tannin–CH_2–tannin

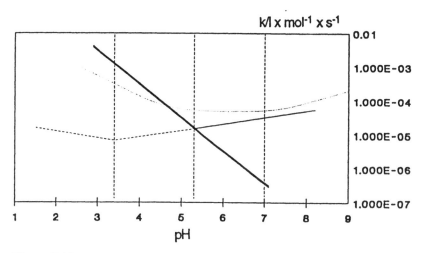

Figure 5.18 Reaction kinetics of mimosa TF and UF resin as a function of pH. Dotted curve, U + F = UF; heavyweight solid curve, U + UF = UMU; lightweight solid curve, tannin.

bridges. Thus, on the microscopic scale, a change from a copolymer to a polymer blend involves a decline in the proportion of mixed methylene bridges. At the point of intersection (pH 5.3) the rates of self-polymerization of the TF and UF resin systems are equal. As there are two self-polymerization reactions and one copolymer reaction with twice the proportion of reaction products, it can be expected that at this pH (5.3) the percentage of copolymers to polymer blends will be approximately 50:50.

Decreasing the pH in the combined resin system to below 5.3 increases the polymer blend percentage over that of the copolymer for the first time. This is because the rate of methylolation at the tannin nuclei is lower than that of the self-condensation of the UF resin. Thus as the pH drops below 5.3 the $-NH-CH_2-NH-$ bridges become more favored than the $-NH-CH_2-$tannin bridges. This percentage increase in the polymer blend cannot be said to tend to about 100%, however, as was the case for the copolymer percentage in the basic region. This is due to the fact that the rate of tannin self-condensation reaction reaches a minimum at pH 3.4 (Fig. 5.18), whereas the pure UF resin system reaction rate decreases exponentially through a pH of 1 to 7. This minimum in the TF curve implies that the self-condensation reaction cannot be taken as negligible as was the case for the UF self-condensation reaction in the alkaline region. If the methylolation reaction is not taken as negligible, a certain amount of copolymer will form and therefore the percentage of polymer blend will not tend to 100%. The actual percentage of polymer blend, however, cannot be determined from these curves, and all that can be said is that at pH 3.4 the proportion of $-NH-CH_2-NH-$ bridges is maximized.

In summary, therefore, region 1, which is any pH above 7, corresponds to a system where the copolymer percentage will tend to 100%. Region 2 (between pH 7 and 5.3) corresponds to a drop from 100% copolymer to 50% copolymer as the pH drops from 7 to 5.3 to yield a final product with a mix of polymer and copolymer. Region 3 (between pH 5.3 and 3.4) corresponds to a further drop in the copolymer percentage, with the percentage of polymer blend reaching a maximum at pH 3.4. The final polymer blend percentage is unknown but is unlikely to tend to 100%. Region 4 (between pH 3.4 and 0.6) corresponds to an increase in the copolymer percentage as the system becomes more acidic. Region 5 (below pH 0.6) corresponds to an increase in the copolymer percentage to its maximum. This maximum is, however, unlikely to exceed 50%, as the rate of methylolation of the tannin nuclei will be similar, if not less than, that of the rate of UF self-polymerization.

B. Fortification with Diisocyanates

Fortification of tannin adhesives with diisocyanates first appeared in 1981 [1,51,52]. It was introduced as a consequence of the then perceived difficulties in utilizing pine tannin extract for particleboard adhesives. This system has now been in very successful commercial operation since 1990 for the manufacture of excellent exterior-grade particleboard [39]. This adhesive system works well not only with pine tannin adhesives, for which it is still used exclusively, but also for all other tannin extracts other than pecan nut tannin (for which it does not offer any technical advantage [4]). It is then of interest to describe the underlying mechanisms and reasons that permit its superior performance. Raw, nonemulsifiable polymeric MDI is the isocyanate used for this application.

In the timber adhesives field, there appears to be a preconception that a nonemulsified MDI cannot be used in water systems, as it reacts with water to form polyureas and is mostly lost to urethane formation with polyols that are present in water solution. The question that should be asked, instead, is at what rate the MDI reacts with the various components of a well-specified water-based adhesive system. In the adhesive systems discussed here, the isocyanate group can react with water, with the phenolic hydroxy group of the polyflavonoid tannins, with the hydroxybenzyl alcohol groups ($-CH_2OH$ methylol groups) formed by the initial reaction of formaldehyde with the phenolic nuclei of the tannins, possibly with the protonated free formaldehyde in the system, and finally with the hydroxy groups of the monomeric and polymeric carbohydrates, which constitute as much as 15% of the tannin extract of commerce. Model compound rates of reaction at different temperatures of MDI with all of these compounds were studied to determine what happens when such an adhesive system is used in practice [39]. The reaction of the MDI with these compounds was followed by FT-IR spectroscopy. Thus, in an adhesive formulation composed of 30 parts MDI, 70 parts pine tannin extract, 7 parts para-formaldehyde, and 85 parts by mass of water, at these molar concentrations, we will have the following relative reaction rates:

MDI/H$_2$O   rate = 1.3×10^{-4} s^{-1}
MDI/catechin   rate = 1.59×10^{-7} s^{-1}
MDI/formaldehyde   rate = 9.4×10^{-7} s^{-1}

From these it would appear at first glance that MDI is, as predicted up to now, deactivated by reaction with water, as the rate of this reaction is several orders of magnitude greater than the reaction of MDI with the phenolic hydroxy-groups of catechin and with the protonated formaldehyde. However, this is not the complete extent of the reactions occurring in the system. The fast reactions of catechin in pine tannins with formaldehyde must also be taken into consideration. Using the rate constant reported previously for this system [53,54], we obtain:

Catechin/formaldehyde to form (catechin– CH$_2$OH)   rate = 1.5×10^{-1} s^{-1}
Catechin–CH$_2$OH/catechin   rate = 5.25×10^{-4} s^{-1}
Catechin–CH$_2$OH/MDI   rate = 3.42×10^{-4} s^{-1}

As the initial reaction of formaldehyde onto catechin to form methylol groups is the faster one, this will be the reaction that dominates the entire system initially. The methylolated catechin, as it forms, can undergo two reactions of almost equal rate. The first is the standard condensation reaction of these phenolic materials to cross-link through –CH$_2$– methylene bridges; the second, just slightly slower, is the attack of the MDI on the hydroxyl of the – CH$_2$OH methylol group formed onto the catechin by the preceding, faster, formaldehyde attack. In such a system the reaction of MDI with water to form polyureas is relegated to fourth position, the latter reaction being slower than the –CH$_2$OH/MDI reaction. These results appear to imply that deactivation of MDI by water is small in such a system, and that urethane formation does indeed occur, a result borne out by the applied bonding results [39].

The mechanism proposed for the total reaction system [39] is:

Reaction Scheme 1

Reaction scheme 2 is as follows:

$$MDI + H_2O \rightarrow R{-}NH{-}\overset{\displaystyle O}{\overset{\displaystyle \|}{C}}{-}NH{-}R$$

In the adhesive system used, with its reagent proportions, by means of the relative rates of reaction, the relative proportions of catechin–formaldehyde polymers, catechin–MDI urethanes and polyurethanes, and of polyureas (the latter by reaction of diisocyanate with water) formed should be, respectively, 53%:35%:12%. Mixed polyurea–catechin species might also be present, although this could not be ascertained. Due to the polymeric nature of the pine tannin species, cross-linking of the polycatechin polymers is likely to occur through both methylene and urethane bridge formation in the cured adhesive. The proportion of catechin–formaldehyde condensates increases in adhesive formulations in which the proportion of MDI decreases in relation to the tannin. The situation is also likely to be quite different in flavonoid tannins presenting a resorcinolic A-ring such as mimosa, rather than a phloroglucinol one such as pine tannin. The reaction rate of mimosa with formaldehyde under the same conditions and subsequent condensation to form mimosa TF polymers are, respectively, 6.94×10^{-3} and 0.480×10^3 L mol^{-1} s^{-1} [53,54].

The same mechanisms as those described will be at play, but the relative proportions of products will then be 61:29:10 for mimosa–formaldehyde, mimosa–urethane, and polyurea, respectively. The adhesive results for mimosa using this system have also been reported to be good [1,39,51,52] but not as good as those for pine tannins. This indicates that it is the final balance of products that appears to determine the strength and water resistance results. Too much urethane might then lead to too much flexibility and poorer timber grip, while too high a proportion of TF condensates to a condition of too high a level of rigidity and brittleness. This type of TF adhesive can be used with fast pressing times, at a high moisture content of the resinated wood particles, and with percentages of MDI as low as 10% of the total adhesive resin solids. The results are always excellent [39].

VII. LOW-FORMALDEHYDE-EMISSION, FAST-PRESSING TF EXTERIOR WOOD ADHESIVES

The combination of high pH (pH 6.6, pecan; pH 7.7, pine), nonmodified or modified, or fortified tannin extracts with 5% paraformaldehyde hardener + 5% urea [12,40], is based on the more recent concept that the faster reacting the tannin, the lower the percentage of formaldehyde needed to cure it to a strong, hard network; the faster the pressing time achievable; and the higher the moisture content that can be tolerated in hot-pressing the board. This increased rate of curing procyanidin–prodelphinidin tannins gives then greater adaptability to a wide range of application conditions. The small percentage of urea added is copolymerized in the network during curing but also functions very effectively as a formaldehyde scavenger during board pressing. The formaldehyde emission in tannin-bonded particleboard is due almost completely to the free formaldehyde residue remaining after pressing, the flavonoid–methylene bonds being impervious to hydrolysis due to their phenolic nature. Thus both the increased reactivity of the tannin and the presence of a urea scavenger contribute to greatly reduce free formaldehyde and hence formaldehyde emission. In the case of pecan nut tannin a small proportion of urea is also necessary to stabilize the tannin [4]. Formaldehyde emission as low as 2 to 4 mg per 100 g of board has been obtained with this system under both laboratory and industrial conditions [40]. The system also works well with the MDI/pine tannin adhesives described earlier [40].

The effectiveness of the 5% paraformaldehyde + 5% urea system led to the suggestion that a system based on a hidden formaldehyde-releasing compound which upon cooling was also capable of trapping the free formaldehyde residue in the board could be equally effective. Thus hexamethylenetetramine (hexamine) was tried as a hardener [40,55]. The hexamine rate of decomposition to formaldehyde and ammonia is greater at acid pH values, thus at the pH values of minimum reactivity of any phenolic materials toward formaldehydes. This is a complicating factor that it was hoped could be overcome by the still rapid reactivity of the procyanidin–prodelphinidin tannins of these polymers. Thus to pecan nut tannin and pine tannin extracts, both adjusted to a pH of 4.5 to 4.6, the former by addition of 2.7% acetic anhydride accelerator [46,40] and the latter at its natural pH (4.5 to 4.6) without any accelerator or with 4% zinc acetate accelerator (predissolved in water) [40], is added a small amount of hexamine predissolved in water. The boards produced with such formulations are excellent with regard to both dry and wet strength as well as presenting very low formaldehyde emission. The results of the V100 test and of the 2-h boil swell test indicated that pecan tannin with 6.5% hexamine gave optimal results and that this formulation performed better with this tannin than with the pine tannin. The higher reactivity of the pecan nut tannin explains this discrepancy between the two tannins. Use of an accelerator such as zinc acetate somewhat improved the results obtained with pine tannin, indicating that the determining factor is the intrinsic reactivity of the tannin with the formaldehyde liberated from the hexamine during hot pressing, but these were still inferior to the pecan nut tannin. Formaldehyde emissions by perforator test in the range 0.2 to 0.6 mg of formaldehyde per 100 g of board were obtained.

The pine + hexamine formulation also suffers from the defect that considerable attention must be given to both the extent of sulfitation of the tannin and, if sulfitation is high, to the relative compression ratios of the board surfaces and core. Thus with this formulation, when the moisture contents of the core and surface are the same, or there is little difference between them, or when the core moisture content (before pressing) is slightly higher than that of the surface, the formulation performs well. In the more traditional pattern of higher surface to lower core moisture contents, the performance of the pine tannin + hexamine formulation is substantially poorer. This is not the case for the pecan nut tannin and hexamine formulation, which performs excellently under all of the foregoing application conditions. As regards the hexamine formulations, a few important considerations must be added. Immediately upon addition of the hexamine solution at ambient temperature, the tannin solution forms a very noticeable soft gel which is then transformed into a colloidal dispersion by energetic mechanical stirring and addition of some water. The viscosity of such a dispersion is low, enabling the adhesive to be applied to the wood particles very easily. Furthermore, the tannin–hexamine dispersion remains stable at ambient temperature for a long period, completely eliminating any problem of pot life. Pine tannin/hexamine water dispersions were measured to be stable, and maintained stable viscosity even 21 days after mixing of the tannin adhesive and hexamine hardener at ambient temperature. This is an important characteristic, as it allows the preparation of stable tannin adhesives in which the hardener is already included, thus of an indefinitely stable "one-step" premixed tannin adhesive glue mix at ambient temperature which is activated only at the higher temperatures used in hot-pressing the particleboard.

Of importance in the case of pine tannins/hexamine formulations is the level of sulfitation of the tannin: the lower the level, the better the result. Hexamine-based formulations are particularly sensitive to the level of sulfitation of the tannin, much more so than when paraformaldehyde is used as a hardener. The use of hexamine in tannin adhesives is not new, having been reported before [1,56]. A mimosa-tannin-based adhesive hardened with hexamine has already been used for interior applications [56,57]. What is new, however, is the capability of the tannin–hexamine system to give excellent V100 results, a characteristic exclusive to the fast-reacting procyanidin (pine) and particularly prodelphinidin (pecan) tannins [40,55]. Previous experiments with the slower-reacting profisetinidin–prorobinetinidin tannins such as mimosa and quebracho have, instead, shown clearly that with the slower tannins it is not possible to obtain boards satisfying the V100 exterior requirements when using hexamine as hardener [55].

It is also important to note that the low formaldehyde emission of boards bonded with tannin–hexamine systems was later found to be due to cross-linking of the tannin–not passing first through decomposition of hexamine to formaldehyde and ammonia [76,77], as believed previously. There is strong evidence that hexamine decomposition to formaldehyde constitutes a minimal part of the mechanisms involved in this type of adhesive formulation. Thus the gel time at 100°C of a tannin extract such as pecan nut that at pH 4.5 is approximately 65 to 70 s when using paraformaldehyde hardener becomes 37 to 40 s with hexamine under the same conditions [77]. A similar trend is also noticeable to a different extent for other tannins; for instance, pine tannin gel times change from 80 to 90 s to 50 to 60 s when using hexamine at pH 4.5. This implies that the cross-linking mechanism does not first pass through decomposition of hexamine to formaldehyde, as in this case the gel time would be slower, not faster [76,77]. It has been shown that attack on the tannin free reactive C6 and C8 sites is carried out by the carbon α to the hexamine nitrogen before decomposition of the hexamine and often without significant hexamine decomposition [76,77]. The mechanisms that were advanced for this behavior also explained the preliminary ambient temperature pregelling observed for tannin–hexamine mixtures [40,55]. Such behavior for hexamine is not unknown, even in the field of synthetic phenol–hexamine resins [78]. ^{13}C NMR studies of PF resins prepared using hexamine exclusively showed the formation at high temperature, during the first 10 min of reaction, of primary, secondary, and tertiary benzylamines and their permanence for well over 1 h of reaction [78]. Thus, in the case of polymeric, highly reactive tannins, where the reaction to cross-linking occurs during the pressing time of the board, thus having a reaction time well below 10 min, little decomposition of the hexamine to formaldehyde occurs [76,77]. This is an important conclusion, as it implies that in the case of tannins, hexamine is not a formaldehyde-yielding compound. The cross-linked structures formed are very stable, as even at the reflux temperature of toluene as used in the standard perforator test for formaldehyde emission, they remain stable enough to yield little or no formaldehyde emission. The higher the reactivity of the tannin, the lower the proportion of hexamine decomposition; this explains the exterior capability of fast tannin–hexamine systems [12,40, 55,76,77] and the effective lack of it for the slower-reacting tannins (such as mimosa and quebracho), in which a greater proportion of decomposition of hexamine to formaldehyde appears to occur [40,56]. This is an important result, as it opens the way for tannin adhesives of low or no formaldehyde emission and of full exterior-grade capability, hence for natural materials of the same performance capability as that of diisocyanates but of no inherent toxicity. It also possibly opens the way to hexamine-based, rather than formaldehyde-based synthetic adhesives if a similar approach is extended.

It is also of interest to note that in a separate field of application—foundry core resins—another system to produce low-emission resins is presently in use. This consists of a polyflavonoid tannin dissolved in furfuryl alcohol to which glyoxal has been added [80]. Although there is not more than 20 to 25% tannin in this system, and the resin is not water carried, it is important to note the use of the aldehyde glyoxal: another route to the elimination of formaldehyde in tannin resins. The system cannot be considered aldehyde-free, however.

VIII. ALDEHYDE-FREE, ZERO-EMISSION, AUTOCONDENSED TANNIN ADHESIVES

While the reactions of the natural polymeric materials with formaldehyde to give polycondensates have been used extensively, autocondensation reactions characteristic of these materials have only recently [58] been used to prepare adhesive polycondensates in the absence of aldehydes. Tannin autocondensations are known, but as they have been studied only on monomeric model compounds, their effect on polymeric systems has neither been studied nor even considered for resin preparation and hardening. Recently, a predominantly prodelphinidin tannin, pecan nut pith tannin extract (*Carya illinoensis*), has been found to undergo fairly rapid autocondensation reactions [4,23,58], although not to such an extent as to lead to cross-linked, hardened resins. This autocondensation reaction is based on the opening under alkaline or acid conditions of the O1–C2 bond of the gallocatechin repeating unit and subsequent autocondensation between the carbocation at C2 of the open unit with the free C6 or C8 sites on a gallocatechin unit of another polymer chain [4,23,58]. Although this reaction leads to noticeable increases in viscosity [4] of concentrated (40%) tannin solutions, gelling does not occur. However, the chance finding that concentrated water solutions of pecan nut tannin extract gelled and hardened in a very short time at alkaline pH by addition of small, even trace amounts of dissolved silica (silicic acid), even at ambient temperature, prompted extensive reevaluation of this approach with surprising and excellent results. The findings obtained on pecan nut (*Carya illinoensis*) pith tannin extract were also tested on pine (*Pinus radiata*) bark extract (a predominately procyanidin tannin), black wattle or mimosa (*Acacia mearnsii* formerly *mollissima*, de Wildt) bark extract (a predominantly prorobinetinidin extract), and quebracho (*Schinopsis balansae*) wood extract (a predominantly profisetinidin extract) [7] with equivalent but not identical results.

Table 5.5 Gel Times as a Function of pH for Commercial Polyflavonoid Tannin Extracts by Autocondensation Induced by 3% SiO_2 (Mass/Mass Tannin Extract)

				Gel time (s)				
pH	Natural pecan	Natural mimosa	Natural pine	Natural quebracho	Semisulfited pine	Semisulfited quebracho	PF	PRF
				Powder SiO_2 (94°C)				
−0.37	—	823	Does not gel	Does not gel	Does not gel	Does not gel	Does not gel	Does not gel
0.40	225	Does not gel	Does not gel	Does not gel	Does not gel	Does not gel	Does not gel	Does not gel
1.0	540	Does not gel	Does not gel	Does not gel	Does not gel	Does not gel	Does not gel	Does not gel
4.22	—	Does not gel	Does not gel	Does not gel	Does not gel	Does not gel	Does not gel	Does not gel
7.69	Does not gel	Does not gel	Does not gel	Does not gel	Does not gel	Does not gel	Does not gel	Does not gel
8.00	85	230	119	752	Does not gel	Does not gel	Does not gel	Does not gel
9.00	60	104	85	—	Does not gel	Does not gel	Does not gel	Does not gel
9.50	—	—	—	110	Does not gel	Does not gel	Does not gel	Does not gel
10.00	52	87	60	98	Does not gel	Does not gel	Does not gel	Does not gel
11.00	43	73	42	73	Does not gel	Does not gel	Does not gel	Does not gel
12.00	35	63	30	—	3900	3600	Does not gel	Does not gel
				Dissolved SiO_2 (25°C)				
12.00	Instantaneous	Instantaneous	Instantaneous	Instantaneous	Does not gel	Does not gel	Does not gel	Does not gel

Source: Ref. 58.

Table 5.6 Gel Times as a Function of % SiO_2 (Mass/Mass Tannin Extract) of Pecan Nut Pith and Mimosa Bark Tannin Extract at Two Comparable pH Values

Silica %	Gel time (s)			
	Mimosa pH 8.29	Mimosa pH 10.60	Pecan pH 8.30	Pecan pH 12.60
1	Does not gel[a]	Does not gel[a]	210	135
2	516	197	92	59
3	241	91	77	50
4	189	71	72	47
6	141	55	65	42

[a]No gel after 3600 s.
Source: Ref. 58.

The trends in gel times at 94° C and ambient temperature (Table 5.5) show that gelation and hardening of 40% water solutions of pecan and other tannin extracts occur readily on addition of small amounts of SiO_2 dissolved in sodium hydroxide. From just these data (Tables 5.5, 5.6) it is not possible to deduce if tannin cross-linking occurs because dissolved silica (or better, silicic acid), acts as a hardener, or because it acts as some sort of catalyst for tannin autocondensation. The percentage of silicic acid used does influence the rate of gelation but over a few percentages not to a great extent, possibly inferring that a catalytic effect is involved (Table 5.6). Nondissolved SiO_2 clearly does not present the same behavior, and the tannin does not harden: the rate of gelation varies according to the pH used (Tables 5.5 and 5.6), this depending somewhat on the solubility, or lack of it, of SiO_2 at different pH values.

When in solution in alkaline environment, monomeric catechin autocondenses readily to give the well-known catechinic acid rearrangement [59]. This is not the case when silicic acid is present for the natural polymer in which a repeating unit is part of a long chain and is thus severely hindered by a number of other repeating units preceding and following it along the same chain [58]. Equally, the acceleration by SiO_2 of heterocycle opening decreases the predominance of interflavonoid cleavage [58]. Such severe hindrance ensures, instead, that the reactive C2 reacts with the free C6 or C8 sites of a flavonoid unit in another chain, leading to cross-linking. Thus the proposed reaction is as follows [58]:

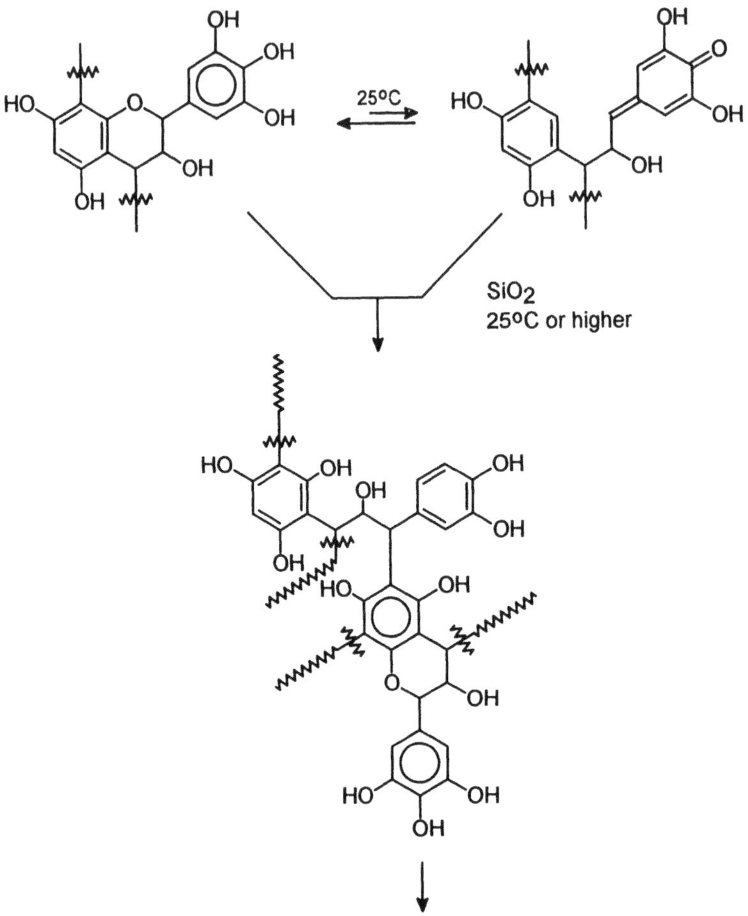
OH
OH
HO
O
OH
OH
OH
25°C
OH
O
HO
OH
OH
OH
SiO2
25°C or higher
HO
OH
OH
OH
OH
OH
HO
OH
O
OH
HO
OH
OH

Why does silicic acid catalyze the autocondensation of the tannin to hardness? The only study in this field indicates that the catalytic activation occurs because silicic acid behaves as a Lewis acid [58]. ^{29}Si NMR has indicated that Si passes through a coordination number of 5 during the reaction because silicon can accept more electrons in its valence shell on the 3d orbital even when the octet on the 3s and 3p orbitals is complete. Furthermore, the possible formation of o-diphenol complexes by Si(OH)$_4$ on the flavonoid unit B-rings is not contributing to the mechanism because these complexes were not detected by potentiometry, ^{13}C NMR, or ^{29}Si NMR [58]: hence the B-ring OH groups are not involved. Equally, the well-documented [60] reaction of phenolic resins hydroxyls with siloxanes to form Si–O–Ar bonds is not involved or is not the determining reaction by itself because such a mechanism alone did not lead to gelation of PF resins (Table 5.5), and thus it would not account by itself for the gelation of the tannin.

It is of interest (Table 5.7) which other materials give gelling of the tannins and which do not. The finding is that it is not the polymeric nature of the acid that determines gelling (Table 5.7), but that the weaker the acid, the faster the gelling. The common thread in Table 5.7 is that all the materials that do gel and harden the tannin by autocondensation behave as Lewis acids. Boric acid ionizes in an unusual way in water: instead of donating one of its

Table 5.7 Gel Time by Autocondensation of Polyflavonoid Pecan Nut Tannin Extract Induced at Various pH Values by Different Lewis Acids

	Gel time (s)			
pH	SiO_2 ($pK_a = 10$)	Boric acid ($pK_a = 9.22$)	$AlCl_3$ ($pK_a = 8.6$)	Phosphoric and polyphosphoric acid (Strong acid)
4.9	Does not gel	5400	—	Does not gel
4.65	Does not gel	2040	—	Does not gel
4.96	Does not gel	1800	—	Does not gel
5.50	Does not gel	1140	—	Does not gel
6.00	Does not gel	280	—	Does not gel
6.53	Does not gel	102	—	Does not gel
7.18	Does not gel	83	—	Does not gel
7.88	112	(270)[a]	—	Does not gel
8.66	60	(360)[a]	—	Does not gel
9.55	49	(360)[a]	780	Does not gel
12.00	35	—	—	Does not gel

[a]Rapid agglutination at ambient temperature causing apparent lengthening of gel time.
Source: Ref. 58.

H^+ atoms to water, it removes an OH^- from a water molecule, leaving an H^+, which combines with water to form H_3O^+ [61]. Thus $B(OH)_3$ also behaves in water as a Lewis acid, the boron atom having a vacant $2p$ orbital that can accept an electron pair from H_2O or from other lone pair-carrying groups [61]. $AlCl_3$ also is well known to behave in this manner as a Lewis acid [47,61]. $AlCl_3$ is also known to be an agent facilitating the cleavage of aromatic ethers [62–65]. In this context, the reaction

has been reported several times [62–65].

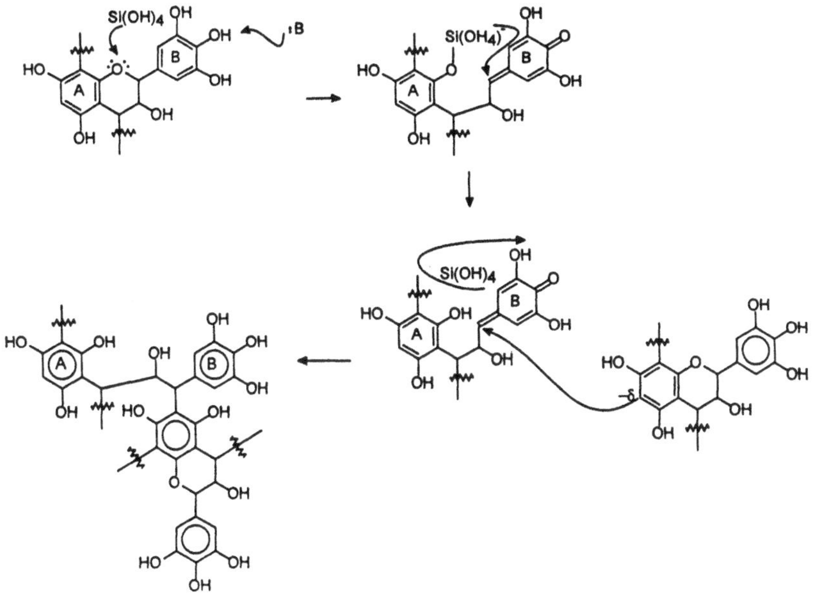

Figure 5.19 Proposed mechanism of Si(OH)$_4$ and other weak Lewis acids catalysis of heterocycle opening and tannin autocondensation, according to accepted pyran ring-opening mechanisms induced by bases.

Mono- and dimeric model compound studies of procyanidins (pine) and profisetinidins (quebracho) tannins have indicated that alkali interflavonoid bond cleavage precedes opening of the pyran ring [3,21,66]. It is now known that in predominately prodelphinidin (pecan) and prorobinetinidin (mimosa) tannins the opening of the pyran heterocycle precedes the inter-flavonoid bond cleavage (if interflavonoid cleavage occurs extensively at all) [4,17,58]: this is due to the greater susceptibility to cleavage of the O1–C2 bond induced by the base deprotonation of the C4' hydroxyl [67] as a consequence of the stronger nucleophilicity of the pyrogallol B-ring. Thus, in procyanidin and profisetinidin tannins, when catalysts such as SiO$_2$ are not present, the reaction in alkali will take a course already reported [68], which does not lead to autocondensation or to increase in molecular mass. Without SiO$_2$, however, prodelphinidin and prorobinetinidin tannin solutions do not gel by themselves, but due to the limited number of units undergoing pyran ring opening and autocondensation in the absence of a catalyst should present a noticeable increase in viscosity just on standing at ambient temperature, or being heated, at high pH: this increase in viscosity has been reported [4,69]. This indicates that in predominantly prodelphinidin (pecan) and prorobinetinidin (mimosa) tannins, gelling is induced by the acceleration or greater extent of pyran ring openings induced by the dissolved SiO_2: the reaction is then highly specific to cleavage of the pyran ring. In predominantly procyanidin (pine) and profisetinidin (quebracho) tannins, the specificity of the acceleration by dissolved SiO_2 for pyran ring cleavage ensures that interflavonoid cleavage does not remain the favorite reaction. If acceleration of pyran ring opening is very marked, interflavonoid bond cleavage will no longer occur, or will occur to a greatly reduced extent. The behavior of these two tannins will now be identical to what is observed for pecan and mimosa (Table 5.5). It is interesting to note that the mechanism that has been proposed (Fig 5.19) would also be consistent with the more recent concept of a one-electron (radical) mechanism being involved in pyran ring cleavage [70].

The reaction can be better explained, however, by a carbocation mechanism [58], because it also occurs at neutral and mildly acid pH values, indicating that base-induced deprotonation of phenolic OH's contributes only to facilitate the reaction at high pH values but is not the determining cause of the reaction. The formation of negative and positive (carbocation) charges would also ensure that no internal rearrangement of the unit to catechinic acid or phlobatannin can occur. This difference from model compound reactions has actually been observed on the natural tannins themselves [58].

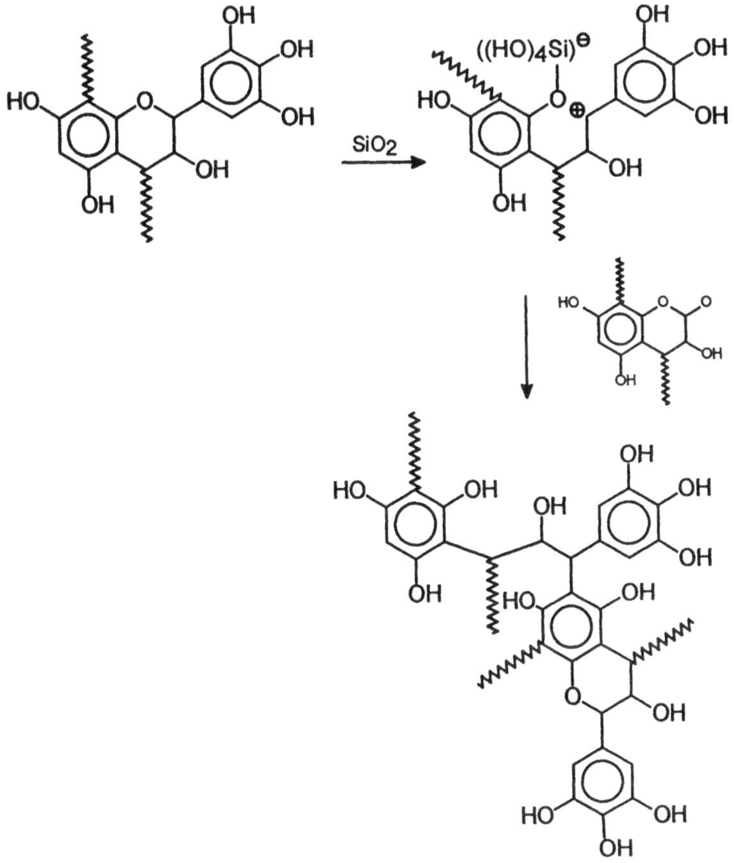

((HO)$_4$Si)$^{\ominus}$

SiO$_2$

A carbocation mechanism would also explain the higher than expected increase in C4–C8/C4–C6 interflavonoid linkages. It is well known that in chain transfer in cationic polymerization a hydride ion can often be extracted by a carbocation from the α-carbon to a double bond or aromatic ring of another molecule [47]. This indicates that it is the additional contribution from the flavonoid chain's lower terminal units that is likely to lead to the appearance and increase in intensity of the C4–C8/C4–C6 ^{13}C NMR band. It also indicates that the counterion–OSi(OH)$_4^-$ charge contributes to peg the unit on the carbocation formed at C2, avoiding bond rotation and eliminating catechinic acid rearrangements.

14

Some confirmatory evidence of Si in a coordination number higher than 4 was obtained not only by ^{29}Si NMR but also by powder x-ray diffraction traces of solid SiO_2, of unreacted powder mixtures of pecan tannin extract and SiO_2 (4% by mass), and of the reacted pecan tannin/SiO_2 (4% by mass) solid network [58]. The hardening system of tannins, in the presence of silicic acid, boric acid, aluminum chloride, or other Lewis acids can be used to bind excellent, zero-aldehyde-emission (as no aldehydes are used to harden the tannin) interior particleboard. Such interior particleboard has better resistance than UF-bonded particleboard to boiling water, but not nearly good enough resistance for exterior-grade application [58]. Internal bonds strengths, dry, for these types of zero-emission interior particleboard are of the order of 0.6 to 0.7 MPa.

There is a further, important point of interest in the application of the autocondensation reactions of tannins to particleboard. Cellulose, in particular crystalline cellulose, also accelerates autocondensation of the tannin [71]. This means that for the tannin that is most prone to such autocondensation behavior—pecan nut tannin—addition of a Lewis acid in the glue mix is not needed. The accelerating surface effect of cellulose is sufficient to give good interior properties to the particleboard when using as the binder a very alkaline solution of pecan tannin extract, or even pine tannin (to a lesser extent), with no Lewis acid or aldehyde added [71,79]. The other tannins, such as mimosa, give the same effect on cellulose, but the strength obtained is much lower, and their gelling and hardening must then be aided by the addition of a Lewis acid such as silicic acid [58,71,79].

A differential scanning calorimetry (DSC) study [71] of the reaction of both SiO_2- and cellulose-induced tannin autocondensation indicated that both reactions of activation are exothermic and that clear exothermic peaks for them could be observed in the DSC trials, and presented energies of activation for them [71]. The autocondensation reaction, for instance, which in pine tannin alone is highly unfavorable at an activation energy of 200 kcal mol^{-1}, becomes much more favorable at 67 kcal mol^{-1} just by the addition of 4% Si(OH)$_4$ and very favorable at 8.5 kcal mol^{-1} for a 1:1 mass ratio of pine tannin extract to cellulose. The reasons that cellulose also catalyzes the reaction of autocondensation of the tannin are similar to those discussed for Lewis acids. Pizzi et al. [72] have shown that cellulose has a surface activation, a heterogeneous catalysis effect on the polymerization and hardening of synthetic PF resins by markedly decreasing the energy of self-condensation of phenolic resins. This occurs because the mass of combined secondary forces binding the molecules of any PF oligomer to the surface of the lignocellulosic substrate is very high [16,72–74], leading to pronounced weakening of the bonds most susceptible to cleavage in the PF oligomer molecule [72]. These bonds are generally the ones that need to be cleaved for further PF polymerization and hardening, thus facilitating and accelerating PF curing. The effect on the autocondensation of the tannins is very similar, cellulose functioning as an activating template for further reaction. A molecular mechanics study [71] has, for instance, shown that in the case of a diflavonoid, a further decrease of 13 kcal mol^{-1} (and 31 kcal mol^{-1} for catechin monomer) will be concentrated on the more labile bonds of the molecule, explaining the catalytic template effect: this type of effect is quite well known in other fields of heterogeneous catalysis [75]. The effect is shown in Fig. 5.20.

It is also important to note that steric hindrance introduced by the strong secondary forces binding a polyflavonoid to cellulose ensures that on heterocycle opening catechinic acid and phlobatannin rearrangements do not occur or are minimal [71]; and that in procyanidin-type tannins such as pine, the respective facility of cleavage of interflavonoid linkage and heterocycle is again inverted, as for the case of Lewis-acid-induced catalysis [71].

IX. FORMULAS

A. MDI/Pine Tannin Adhesive for Particleboard [39]

To 60 L of 40% pine tannin extract solution are added 8.5 L of polymeric raw MDI (Bayer 44 V20 or VK), 6.0 L of water, 3.5 kg of wax emulsion,

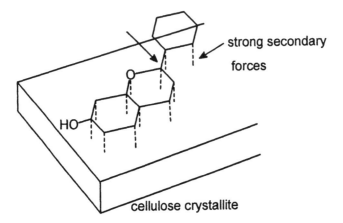

strong secondary
forces

HO

cellulose crystallite

Figure 5.20 Schematic catalytic effect induced by a cellulose surface on reactive bonds weakening of a flavonoid.

and 3.5 kg of paraformaldehyde fine powder 96 to 97% grade in the glue mix for the particleboard core. The surface glue mix is composed of 60 L of 40% pine tannin extract water solution, to which are added 2.0 L of polymeric raw MDI, 5.0 to 6.0 L of wax emulsion, and 3.5 kg of paraformaldehyde. The percentage adhesive solids (calculated as the sum of tannin extract solids + MDI) on dry wood used are of 10% in the board core and 13 to 14% on board surfaces. Low-formaldehyde-emission panels can be obtained substituting for the 3.5 kg of paraformaldehyde, 1.75 kg of paraformaldehyde, and 4.4 kg of a 40% water solution of urea, the paraformaldehyde and urea solution being added separately to the glue mix.

B. Traditional Mimosa Extract Particleboard Adhesives [37]

To 100 parts by mass of a 55% water solution of mimosa tannin extract are added 0.15 parts of industrial defoamer, at ambient temperature. The reaction mix is heated to 80°C under mechanical stirring and 2 parts by mass of acetic anhydride or maleic anhydride is added, followed between 5 and 15 min later by 2 parts of phenyl acetate (optional). The temperature is raised to 90 to 93°C 45 min after and enough 30% NaOH solution is added as rapidly as possible to reach a pH of 7.8 to 8.0. The mixture is then left to digest for 3 h but not more than $3\frac{1}{2}$ h under mechanical stirring at 90 to 93°C. The mixture is then cooled as rapidly as possible and the pH adjusted with acetic acid to the required pH of reactivity (pH 7.0 for traditional applications, much higher pH for fast-pressing applications). This adhesive intermediate diluted to the wanted viscosity has added to it 8 to 10% paraformaldehyde fine powder just before addition to the wood particles.

REFERENCES

1. A. Pizzi, Chapter 4 in *Wood Adhesives: Chemistry and Technology*, Vol. 1 (A. Pizzi, ed.), Marcel Dekker, New York, 1983.
2. A. Pizzi, *J. Macromol. Sci. Rev.*, *C18*(2): 147 (1980).
3. R. M. Hemingway, P. E. Laks, G. W. McGraw, and R. E. Kreibich, in *Wood Adhesives in 1985: Status and Needs*, Forest Products Research Society, Madison, Wis., 1986.
4. A. Pizzi and A. Stephanou, *Holzforsch. Holzverwert.*, *45*(2): 30 (1993); *J. Appl. Polym. Sci.*, in press (1993).
5. G. W. McGraw, T. G. Rials, J. P. Steynberg, and R. W. Hemingway, in *Plant Polyphenols* (R. W. Hemingway and P. E. Laks, eds.), Plenum Press, New York, 1992.
6. D. G. Roux, Chapter 16 in *Adhesives from Renewable Resources* (R. W. Hemingway and A. H. Conner, eds.), ACS Symposium Series No. 385, Washington, D.C., 1989.
7. A. Pizzi and A. Stephanou, *Holz Roh Werkst.*, April (1994).
8. A. Pizzi and A. Stephanou, *J. Appl. Polym. Sci.*, in press (1993).
9. D. Ferreira, J. P. Steynberg, J. F. W. Burger, and B. C. B. Bezuidenhoudt, in *Plant Polyphenols* (R. W. Hemingway and P. Laks, eds.), Plenum Press, New York, 1992.
10. R. E. Kreibich, R. W. Hemingway, and W. T. Nearn, *For. Prod. J.*, *43*(7/8): 45 (1993).
11. A. Pizzi, D. du T. Rossouw, W. E. Knuffel, and M. Singmin, *Holzforsch. Holzverwert.*, *32*(6): 140 (1980).
12. A. Pizzi and Bakelite A. G., unpublished results, 1993.
13. S. Ohara and R. W. Hemingway, *J. Wood Chem. Technol.*, *11*(2): 195 (1991).
14. V. J. Sealy-Fisher and A. Pizzi, *Holz Roh Werkst.*, *50*: 217 (1992).
15. T. Moledi, B.Sc.(Hons) research project, University of the Witwatersrand, Johannesburg, South Africa, 1990.
16. A. Pizzi and G. de Sousa, *Chem. Phys.*, *164*, 203 (1992).
17. A. Pizzi and A. Stephanou, *J. Appl. Polym. Sci.*, *50*: 2105 (1993).
18. R. Newman and L. J. Porter, in *Plant Polyphenols* (R. W. Hemingway and P. E. Laks, eds.), Plenum Press, New York, 1992.
19. Leather Industries Research Institute, *Modern Applications of Mimosa Extract*, Liri, Grahamstown, South Africa, 1965, pp. 33–51.
20. I. Abe, M. Funaoka, and M. Kodama, *Mokuzai Gakkaishi*, *33*: 582 (1987).
21. R. W. Hemingway and G. W. McGraw, *J. Wood Chem. Technol.*, *3*(4): 421 (1983).

22. G. W. McGraw, T. G. Rials, J. P. Steynberg, and R. W. Hemingway, in *Plant Polyphenols* (R. W. Hemingway and P. E. Laks, eds.), Plenum Press, New York, 1992.

23. A. Pizzi and A. Stephanou, *J. Appl. Polym. Sci.*, in press (1993).

24. D. Thompson, M.Sc. thesis, University of the Witwatersrand, Johannesburg, South Africa, 1994.

25. D. Thompson and A. Pizzi, *J. Appl. Polym. Sci.*, in press (1994).

26. A. Pizzi, unpublished work, 1974.

27. E. von Leyser and A. Pizzi, *Holz Roh Werkst.*, *48*: 25 (1990).

28. J. Baeza, private communication, 1990.

29. L. Foo and J. J. Karchesy, in *Chemistry and Significance of Condensed Tannins* (R. W. Hemingway and J. J. Karchesy, eds.), Plenum Press, New York, 1989.

30. D. du T. Rossouw, *Looistofinhoud en-gehalte van Suid Afrikaanse dennebos*, CSIR special project HOUT 163, Pretoria, South Africa, 1978.

31. D. du T. Rossouw, *Ontwikkeling van dennelooistoflym*, CSIR special report HOUT 137, Pretoria, South Africa, 1978.

32. H. A. Schroeder, *Pine Tannin Adhesives*, CSIR special report HOUT 120, Pretoria, South Africa, 1976.

33. K. Freudenberg and J. Alonso de Lama, *Annalen*, *612*: 78 (1958).

34. R. W. Hemingway and L. Y. Foo, *J. Chem. Soc. Chem. Commun.*, 1035 (1983).

35. R. Brown and W. Cummings, *J. Chem. Soc.*, 4302 (1958).

36. R. Brown, W. Cummings, and J. Newbould, *J. Chem. Soc.*, 3677 (1961).

37. A. Pizzi, *For. Prod. J.*, *28*(12): 42 (1978).

38. A. Pizzi, *Holzforsch. Holzverwert.*, *43*: 83 (1991).

39. A. Pizzi, E. P. von Leyser, J. Valenzuela, and J. G. Clark, *Holzforschung*, *47*: 164 (1993).

40. A. Pizzi, J. Valenzuela, and C. Westermeyer, *Holz Roh Werkst.*, in press (1993).

41. Spanplatte: Flach-pressplatten für das Bauwesen, DIN 68763 (1982).

42. J. van Niekerk and A. Pizzi, *Holz Roh Werkst.*, Jan–Feb (1994).

43. A. Pizzi, *Holzforsch. Holzverwert.*, *31*: 85 (1979).

44. A. Huber, private communication, 1993.

45. A. Pizzi and A. Stephanou, *J. Appl. Polym. Sci.*, *49*: 2157 (1993); *Holzforschung*, *48*: 150 (1994).

46. A. Pizzi and A. Stephanou, *J. Appl. Polym. Sci.*, *51*: 1351 (1994).

47. H. R. Allcock and F. W. Lampe, *Contemporary Polymer Chemistry*, 2nd ed., Prentice Hall, Englewood Cliffs, N.J., 1990.

48. A. Pizzi, *Adhesives Age*, *20*(12): 27 (1977).

49. P. A. J. L. Custers, R. Rushbrook, A. Pizzi, and C. J. Knauff, *Holzforsch. Holzverwert.*, *31*: 6 (1979).

50. N. Meikleham and A. Pizzi, *J. Appl. Polym. Sci.*, in press (1993).

51. A. Pizzi, *Int. J. Adhes. Adhes.*, 2: 261 (1981).

52. A. Pizzi, *Holz Roh Werkst.*, *40*: 293 (1982).

53. D. du t. Rossouw, A. Pizzi, and G. McGillivray, *J. Polym. Sci. Chem. Ed.*, *18*: 3323 (1980).
54. A. Pizzi and P. van der Spuy, *J. Polym. Sci Chem. Ed.*, *18*: 3447 (1980).
55. A. Pizzi, B. Dombo, and W. Roll German patent application (1993).
56. A. Pizzi, Ph.D. thesis, University of the Orange Free State, Bloemfontein, South Africa, 1977.
57. A. Pizzi, unpublished work, 1974.
58. N. Meikleham, A. Pizzi, and A. Stephanou, *J. Appl. Polym. Sci.*, in press (1994).
59. S. Ohara and R. W. Hemingway, *J. Wood Chem. Technol.*, *11*(2): 195 (1991).
60. A. Knop and L. A. Pilato, *Phenolic Resins*, Springer-Verlag, Berlin, 1985.
61. R. J. Gillespie, D. A. Humphreys, and N. C. Baird, E. A. Robinson, Chemistry, Allyn and Bacon, Boston, 1986.
62. C. Weygand, *Organic Preparations*, Interscience, New York, 1945, p. 196.
63. G. Hartmann and B. Gattermann, *Ber.*, *25*: 3531 (1892).
64. P. Pfeiffer and W. Loewe, *J. Prakt. Chem.*, *147*: 293 (1936).
65. P. Pfeiffer and H. Hasack, *Annalen*, *460*: 156 (1928).
66. P. E. Laks and R. W. Hemingway, *J. Chem. Soc. Perkin Trans I*, 465 (1987).
67. W. B. Whalley and P. P. Mehta, *J. Chem. Soc.*, 5327 (1963).
68. P. E. Laks and R. W. Hemingway, *Holzforschung*, *42*: 287 (1987).
69. A. Pizzi, *Int. J. Adhes. Adhes.*, *1*: 107 (1980).
70. J. A. Kennedy, M. H. G. Munro, H. K. J. Powell, L. J. Porter, and L. Y. Foo, *Aust. J. Chem.*, *37*: 885 (1984).
71. A. Pizzi, N. Meikleham, and A. Stephanou, *J. Appl. Polym. Sci.*, in press (1994).
72. A. Pizzi, B. Mtsweni, and W. Parsons, *J. Appl. Polym. Sci.*, in press (1993).
73. A. Pizzi and N. J. Eaton, *J. Adhes. Sci. Technol.*, *1*(3): 191 (1987).
74. A. Pizzi and S. Maboka, *J. Adhes. Sci. Technol*, *7*(2): 81 (1993).
75. G. C. Bond, *Heterogeneous Catalysis: Principles and Application*, Oxford Science Publishers, Clarendon Press, Oxford, 1987.
76. T. Stanbury, B. Sc.(Hons). project, University of the Witwatersrand, Johannesburg, South Africa, 1993.
77. A. Pizzi, *Holz Roh Werkst.*, Kurz Originalia in press (1994).
78. A. Sojka, R. A. Wolfe, and G. D. Guenther, *Macromolecules*, 1539 (1981).
79. A. Pizzi, B. Dombo, and W. Roll, German patent application (1993).
80. W. McKillip, private communication, 1994.

6

Lignin-Based Wood Adhesives

I. INTRODUCTION

The occurrence of lignin as a waste product in pulp mills has made it an attractive raw material for adhesives ever since the beginning of the sulfite pulping of wood. The first patents dealing with the application of spent sulfite liquor (SSL) as an adhesive for paper, wood, and other lignocellulosic materials date back to the end of the nineteenth century [1], and since then the literature has continued to grow in number. On the other hand, the technical utilization of lignin on a large scale is still at a very low level for the amount produced worldwide. At present, most of the spent liquor in pulp mills is burned. Only about 20% is used for various purposes, such as dispersants, oil-well-drilling muds, pelletizing materials, molding stabilizers, and concrete grinding additives. As a major wood component, native lignin is neither hygroscopic nor soluble in water. However, during technical sulfite pulping, lignin becomes soluble in water, due to partial degradation and introduction of sulfonic acid groups ($-SO_3H$). To apply SSL as an adhesive, it must be converted to an insoluble state during the curing period.

Cross-linking in lignin can be achieved either by condensation or by radical coupling reactions. A large number of patents have become known during the past three decades [2] dealing with the application of SSL as a wood

219

adhesive, in which the lignin is cross-linked by condensation reactions. However, either high temperatures and long heating times or mineral acids are required for these condensation reactions, which cause structural changes or charring in the wood particles. Recently, cross-linking of the lignosulfonate molecules by radical combinations, which avoids mineral acids and high temperatures, has been developed, but this presents disadvantages as well, as the use of peroxides is not favored in wood-processing plants, for a variety of reasons. The use of lignins by polycondensation reaction with formaldehyde also presents the disadvantage of slower pressing time in their application to panel products.

II. CHEMICAL BACKGROUNDS OF THE CURING REACTIONS OF LIGNIN

Lignin is composed of phenylpropane (C9) units that are linked together by carbon-to-carbon as well as carbon-to-oxygen (ether) bonds. Our present knowledge of lignin structure is based on the assumption that it is formed from p-hydroxycinnamyl alcohols by oxidative coupling [2,3] oxidized by hydrogen peroxide and peroxidase to a phenoxy radical (R). The unpaired electron in R is delocalized and reacts at three different sites of the radical.

Phenyl propanoid units of lignin:
R, R2 = H, OCH_3, R3 = H, CH_3, CH_2
or / and $\left(\frac{\xi}{\xi}\right)$ = possible linkage to
other phenyl propanoid units

A. Cross-Linking by Condensation Reactions

When lignosulfonate is treated with strong mineral acids at elevated temperatures or heated at temperatures above 180°C, condensation reactions leading to diphenyl methanes and sulfones take place. The reactivity of lignosulfonates depends to some extent on the cation. Of the four technically otained lignosulfonates, the calcium-based exhibit the lowest and the ammonium-based the highest reactivity, while the sodium and magnesium lignosulfonates show medium reactivity. Hydroxybenzyl alcohol groups as well as sulfonic acid groups on the carbon α to the aromatic rings of some of the phenyl propane units of the random polymer react with the aromatic nuclei of other phenylpropane units in the presence of strong mineral acids. This reaction, leading to diphenylmethanes, is of the same type as the formation of phenolic resins from phenol and formaldehyde. Lignin also reacts with formaldehyde and can be cross-linked by it, in the same manner of synthetic polyphenolic resins.

B. Cross-Linking by Oxidative Coupling [2]

Lignosulfonic acid in technical SSL contains about 0.4 free phenolic hydroxy group per C9 unit. Therefore, like the formation of lignin in plants, cross-linking of lignosulfonate is possible by oxidative coupling. Oxidants such as hydrogen peroxide, and catalysts such as sulfur dioxide or potassium ferricyanide, are most effective. Treatment of a 50% technical SSL with this redox system leads to a very vigorous exothermic reaction and evaporation of water. The yield of the resin under certain conditions exceeds 70%, indicating that some carbohydrates must also have been enclosed in the resin. The advantage of this type of cross-linking over that of condensation reactions is that it needs neither mineral acids nor high temperatures, due to the recombination of radicals, for which the activation energy is very low. The strongly exothermic reaction causes a uniform temperature profile during pressing of particleboard without external heat.

III. APPLICATION OF LIGNIN AS AN ADHESIVE FOR PARTICLEBOARD, PLYWOOD, AND FIBERBOARD

According to its structure as a polyphenol, lignin as an adhesive should be similar to PF resins. This is true for native lignin in wood, while technical lignins (lignosulfonate and black liquor) have to be additionally cross-linked to change them into insoluble resins. However, condensation reactions in lignin by heat or mineral acids cannot be as effective as in synthetic PF resins, due to the lower number of free positions in the aromatic nuclei of lignin and their considerably lower reactivity than in PF resins. First, there is only 0.5 free o-position (ortho to the phenolic groups) per C9 unit; the 6- and 2-positions are less reactive. Second, there is less than one benzyl alcohol or ether group per C9 unit in lignin, while in synthetic PF resins up to three methylol groups can be introduced into one phenolic ring. Finally, the aromatic nuclei in lignin are considerably less reactive than phenol toward hydroxybenzyl alcohol groups, due to the presence of methoxy or methoxy-equivalent groups rather than hydroxy groups on the lignin aromatic rings. For these reasons, lignin in technical spent liquors cannot be as effectively cross-linked as synthetic PF resins. At a minimum, higher press temperatures at longer heating times or higher acid concentrations are necessary.

Quite a number of patents have been pending during the past three decades dealing with lignin as adhesive for particleboard, plywood, and fiberboard in the absence of conventional PF or UF adhesives [2]. Besides lignin, in most cases additional cross-linking agents for lignin are necessary, such as epoxides, polyisocyanates, polyols, polyacrylamides, polyethyleneimine, al-dehydes, maleic anhydride, amines, proteins, melamine, hydrazine, and so on. So far, these procedures, for different reasons, have not led to any major practical applications. Very few procedures use lignosulfonates or SSL without integrated cross-linking chemicals, like those of Pelikan et al. (1954), Pedersen and Jul-Rasmussen (1963), Shen (1973), Nimz et al. (1972), and others [2]. The patent of Pelikan et al. describes procedures for using lignin as an adhesive for floor tiles by cross-linking it oxidatively with chromium trioxide. The mechanism of cross-linking is the same as with hydrogen peroxide (Nimz, 1972) [2], but it is much less effective. Its applicability to particleboards has been tried, but the boards exhibited low tensile strengths and disintegrated in water at 20°C in less than 2 h.

A. Curing Lignin Boards by Long Pressing Time and Postheating Treatment

According to Pedersen and Jul-Rasmussen [2], wood chips are mixed with 20 to 25% of their weight with a 50% technical SSL and pressed at 185°C for 30 min, giving a 12-mm-thick board that has to be postheated at 195°C for 80 min in an autoclave. The pH value of the SSL had been adjusted to 3 by citric acid. The particleboard obtained had a bending strength of 230 kP cm^{-2}, a tensile strength perpendicular to the grain of 5.3 kP cm^{-2}, and a density of 0.7 g cm^{-3}. Press temperatures may vary between 170 and 235°C and temperatures of the autoclave between 170 and 210°C.

High pressing and autoclave temperatures as well as long heating times are necessary for effective cross-linking by condensation reactions in lignin, as pointed out above. The color of the boards is dark, due to decomposition reactions and charring, caused by the high temperatures, and the density of the boards usually reaches values at around 0.8 g cm^{-3} if the required tensile strength is to be obtained. The temperature in the core layer during pressing reaches 140°C. This may also cause condensation reactions between wood and SSL as well as chemical and physical changes in the wood particles [5].

The relatively high dimensional stability of the particleboard toward water may be caused by these changes. Pedersen and Jul-Rasmussen (1963) found a thickness expansion after a 2-h soaking in water at 20°C of only 1.5% and 13.8% water absorbance. Open-air tests, extending over 5 years, carried out by the wood panel products laboratory of the Technical Research Centre of Finland [6], revealed that SSL boards, obtained according to the Pedersen procedure, were superior in strength and in surface properties to UF as well as to PF particleboard. Roffael [7] has shown that water absorption of Pedersen SSL particleboard at different air humidities is only about half as high as with conventional PF particleboard. Weathering for 1 year gave a nearly constant humidity at around 6% for SSL particleboard, while PF boards gave humidities between 12 and 15%. Also, after soaking the boards in water at 20°C for 24 h, the LS boards lost only 25% of their initial tensile strength, whereas that of conventional PF particleboard decreased by 70%. In contrast, the mechanical strength properties of SSL particleboard were inferior to those of PF boards [8].

The Pedersen procedure has been applied in mill-scale tests in Denmark, Switzerland, and Finland, but has been discontinued in all cases. One reason for this failure is the high cost caused by the two-stage heating treatment. The autoclave must consist of refined steel, due to the evolution of corrosive gases such as sulfur dioxide, causing additional high costs. Other reasons were the long pressing and curing times needed for manufacture. However, one of the main reasons for discontinuation of the procedure were the frequent fires induced by the high pressing and posttreatment temperatures [2].

As mentioned above, the condensation rate of lignosulfonates depends on the cation, with ammonium ions exhibiting the highest reactivity. Shen and Calvé [4] used fractionated ammonium-based SSL as binder for particleboard and found the highest reactivity, leading to the best mechanical board properties, with a low-molecular-weight fraction. Unexpectedly, the tensile strength of dry particleboard obtained with a low-molecular-weight ammonium-based SSL fraction increased with the sugar content of the SSL. Best board properties were obtained with 6% of a low-molecular-weight (0 to 5000) ammonium-based SSL fraction having 50 to 60% sugar. In this case, a pressing time of 8 min at 210°C was sufficient for manufacture of 11-mm-thick waferboard to meet the Canadian standard requirements of exterior-grade particleboard.

Obviously, the sugars take part in the condensation reactions of lignosulfonate by production of fufural. While the bending strength of dry boards increased steadily, with the sugar content of the SSL going up to 80%, the bending strength, after a 2-h boiling of the boards reached a maximum at about 50 to 60% carbohydrates, indicating that the condensation between lignin and carbohydrates leads to better water resistance than that between carbohydrates only.

B. Curing Lignosulfonate Particleboard with Sulfuric Acid

In 1973, K. C. Shen of the Eastern Forest Products Laboratory in Ottawa, Canada, proposed sulfuric acid as a curing agent for SSL waferboard. The pressing conditions were the same as that of conventional PF particleboard, when poplar wafers were first sprayed with 1% of a 15 to 20% sulfuric acid solution and then with 4 to 5% SSL powder, which adheres to the surface of the wet wood wafers. Later [9], concentrated sulfuric acid (9%) was added to the SSL before spray drying, and the powder adhered to the wax-coated wafers. High pressing temperatures of about 205°C were also necessary, the catalytic effect of the sulfuric acid merely reducing the pressing time to that of industrial conditions for PF particleboard.

The strength properties of the boards, having an average specific gravity of 0.67 g cm^{-3}, were measured by the torsion shear at the center plane of 1 × 1 in. specimens, from which the internal bond strength (tensile strength perpendicular to the surface) can be obtained by multiplication with the factor 0.7 [10]. Values obtained for internal bond strength and modulus of rupture (MOR) for dry samples as well as after 2 h of boiling met the Canadian standards for particleboard [11].

The torsion shear strength and MOR of dry boards were independent of the pressing time, while the wet strength increased proportionately with the pressing time. This means that for exterior-grade requirements, distinct pressing times are necessary. Best board properties were obtained with 1% concentrated sulfuric acid, based on dry wood particles. At higher acid concentrations the strength of dry boards decreased, while that of wet boards showed a further increase. However, charring of the wood particles takes place at acid concentrations higher than 0.9%. The thickness expansion of SSL boards obtained with 1% sulfuric acid lay between 26 and 46%, after soaking in water at 20°C for 1 week, and between 51 and 66% after 2 h of boiling. These values are considerably worse than those of exterior-grade PF particleboard.

The acidity of the particleboard was found to be pH 3, after disintegration of the boards in 10 times their weight of water [12]. It has been reported that the acidity had no longer-term influence on the mechanical board properties, checked by conventional accelerated aging treatments [12]. In this case, 11-mm boards had been pressed for 6 min and a part of them had been postheated at 149°C for 2 h. Of the accelerated aging treatments only 20 days of heating at 149°C showed faster aging of SSL than of PF boards, which is due to the higher acidity of the SSL boards.

On the whole, the SSL particleboard obtained by Shen cannot be compared technically with exterior-grade PF particleboard. The Shen procedure, however, has found no practical application yet, as the results obtained are still far from those obtainable with PF particleboard. In 1977, Shen reported that a short production trial run had been carried out at a waferboard plant; although the preliminary results were promising, additional work was still required to modify the binder formulation and production parameters to meet the requirements of plant operation and to obtain results comparable to those of PF particleboard.

C. Curing SSL Boards with Hydrogen Peroxide

The drawbacks inherent in the Pedersen and Shen procedures—high pressing temperatures and long pressing times or strong mineral acids—can be avoided if cross-linking of the lignin molecules is achieved by radical coupling instead of by condensation reactions (Nimz et al. 1972 [2]). In this case, the formation of new carbon–carbon as well as carbon–oxygen bonds between two radicals is a very fast reaction with a low activation energy, which needs no external heating or mineral acids as catalyst. This means that the reaction is very specific, and side reactions such as decompositions and charring can be avoided, while linkages between wood and SSL may also occur.

The essential radicals are formed from phenolic groups in the lignosulfonate molecules by oxidation with hydrogen peroxide in the presence of a catalyst. Out of a number of catalysts, sulfur dioxide (SO_2) has been proven to be the most effective [13]. A 50% calcium-based SSL containing about 1% SO_2 at pH 2 reacts vehemently with a 35% hydogen peroxide solution in a strongly exothermic reaction, forming an insoluble gel. The reaction time is less than 1 min but depends on the source and composition of SSL. At higher pH values, the reaction takes some minutes or needs heating up to about 70°C, but after reaching 70°C, the reaction is also very fast.

It has been reported [13] that the SSL, containing the catalyst, and the hydogen peroxide solution have to be sprayed separately on the wood chips.

Under certain conditions, the hydrogen peroxide can be mixed with the SSL and sprayed together on the wood chips in a single operation [13]. Another possibility consists of adding half of the SSL as spray-dried powder and lowering the humidity of the blended chips to about 13%, which is the upper limit according to German standards. The powder may either be mixed together with the liquid SSL and the hydrogen peroxide or added separately after the wood chips have been sprayed with the SSL solution/hydrogen peroxide mixture. The humidity of the wood chips can thus be adjusted to predetermined values. The pot life of the blended wood chips, which is the assembly time between spraying and pressing, would then be extended. Medium-density interior-grade particleboard can be obtained from wood chips with 20% SSL, based as dry material on dry wood chips, at pressing temperatures between 100 and 120°C under otherwise conventional manufacturing conditions for UF particleboard.

The reasons that this system has not found industrial favor are several: (1) the unfavorable situation due to the presence of a peroxide in wood panel plants, such as possible machinery corrosion, and other problems, and (2) the fact that the board produced is often relatively very soft immediately out of the particleboard press, rendering particularly problematic its early handling.

IV. LIGNIN IN COMBINATION WITH PF ADHESIVES

The number of patents pending on lignin as a substitute or extender for phenolic wood adhesives during the past three decades is high [2]. Under certain conditions, up to 40% of PF adhesive can be replaced by lignosulfonate or black liquor without significantly extending the curing time or worsening board properties. Lignin–PF formulations have been used in manufacturing particleboard, fiberboard, and plywood. The reason for their application has to be seen in the lowering of costs, resulting from the difference in cost between PF and lignin. However, in most cases the lignin has to be pretreated by deionization, ultrafiltration, or cation change. Two recent procedures that have become better known are discussed next in more detail.

A. Lignin–PF Formulations

In 1971, Roffael and Rauch [14,15] claimed that the curing time of SSL particleboard could be reduced and, according to Pedersen, the postheating treatment in an autoclave avoided when phenolic resins of the novolak type were added to the SSL [15]. Due to coagulations between calcium-based SSL and PF, the calcium-based SSL has to be transferred into sodium-based SSL. The board properties are strongly dependent on the pH value of the glue: for instance, 10% SSL, 4% novolak, and 2.1% hexamethylenetetramine were applied to dry pine wood chips to prepare 9-mm-thick boards at a pressing time of 12 min.

While the highest bending and tensile strengths were obtained between pH 5 and 7, the percentage swelling in water at 20°C, after 24 h, had a minimum at pH 3.5. For this reason, a pH value of 4.7 has been suggested by the authors as a compromise [15]. Both the mechanical strength and the dimensional stability of the particleboard can be improved by higher ratios of novolak in the glue formulation or increasing pressing temperatures up to 250°C. Besides conventional contact heating in a flat press, high-frequency heating was applied, raising the temperatures during pressing in the core layer to 220°C for 1 min, which diminished the pressing time. In contrast to their publication in 1971 [15], Roffael and Rauch in 1973 [16] found that phenolic resins of the resol type also improve the binding properties of SSL in particleboard, and the percentage swelling can be improved to meet the German standard specification [18] (6% after 2 h in water at 20°C) by applying a postheating treatment at 200°C for 1 h.

The postheating treatment could be avoided when higher amounts of resol-type resin were used. In conventional PF particleboard the PF resin amounts to about 8%, based on dry wood particles. It has been found [16] that up to 33% of the resol-type adhesive in conventional PF particleboard can be substituted in the surface layers of a three-layer 22-mm board by sodium-based lignosulfonate without major deterioration in the mechanical board properties. In 20-mm one-layer particleboards at pH 9, up to 25% of the PF resin could be replaced by sodium–lignosulfonate under conventional pressing conditions, leading to particleboard meeting the German standards specification [7]. Furthermore, 10% of the PF resin in beech plywood could be substituted for by sodium-based lignosulfonate at a pressing temperature of 165°C, and up to 30% at 190°C [8]. The highest plywood shear strength was obtained with an adhesive formulation of pH 12 to 13.

B. Karatex Adhesive

According to Forss and Fuhrmann (1972) [6,17], the amount of lignin in lignin–PF adhesives for particleboard, plywood, and fiberboard can be increased to 40 to 70% if a high-molecular-weight fraction (molecular weight

> 5000) of either lignosulfonate or black liquor, obtained from alkaline pulping of wood, is applied. Fractionation of SSL or black liquor can be achieved by ultrafiltration [6,17]. According to the authors, the higher effectivity of high-molecular-weight lignin molecules is due to their higher level of cross-linking, which requires less pH for the formation of an insoluble copolymer than do low-molecular-weight lignin molecules. However, bearing in mind the findings of Shen [4] that low-molecular-weight ammonium-based SSL is more effective, Forss and Fuhrmann appear not to have checked the influence of inorganic salts in SSL or black liquor that are separated off during ultrafiltration [2]. Forss and Fuhrmann assume that condensates between smaller lignin molecules and PF "are unable to contribute to the three-dimensional network" [6], which is unlikely because low-molecular-weight lignin molecules are more reactive than are high-molecular-weight molecules.

In the manufacture of particleboard, either high-frequency (HF) heating or combined contact/HF heating has been applied. In the latter case, the press platen temperature was 180°C. German standard requirements for weather-resistant particleboards were met at pressing times between 10 and 12 s mm^{-1} and 8 to 12% adhesive, based on dry wood particles.

One advantage inherent in the fractionation by ultrafiltration is that the lignin becomes more uniform and less dependent on variations in pulping conditions and wood source, which sometimes cause serious problems in the application of technical lignins. Full-scale plywood mill tests, some of them running continuously for several weeks, appear to have been performed in two Finish plywood mills. Again, this does not appear to be in operation anymore.

C. Methylolated Lignins

The fundamental problem of lignin, the slowing of the pressing time obtainable with PF resins was partly eliminated by Sellers (1990) [19] and by Calvé (1990) [20], who first reacted lignin with formaldehyde in a reactor for a few hours. A methylolated lignin (ML) equivalent to a PF resol, was obtained. As in this case the reactivity of the introduced methylol groups of lignin depends on the reactivity of phenolic nuclei available for reaction, mixing with a synthetic PF resin ensures that the reactivity of the PF resin is not impaired. In this manner up to 30% methylolated lignin could be used to substitute the PF adhesive, with no drop in performance or pressing times. Both Sellers and Calvé conducted plywood industrial plant trials with such PF–ML systems obtaining excellent results. It is believed that at least one

Canadian plywood mill is using such a system industrially today. As plywood pressing time is not the really critical variable in a plywood mill, this system did not prove, by itself, suitable for application to particleboard mills, where the shortness of the pressing time that can be obtained is the determining variable.

Attempts were made to use more reactive lignins, such as bagasse (sugarcane waste) lignins, which present 0.7 to 0.9 reactive positions for phenyl propane unit, using the same approach. While good particleboard could be obtained with a mixture of 67% methylolated bagasse lignin and 33% PF resin, these could be obtained only at 37 to 50 s mm^{-1} pressing times, still far too long to be of any interest to a particleboard mill [21]. The low reactivity of lignins toward formaldehyde and the limited number of sites available for reaction with formaldehyde on most aromatic nuclei of the phenyl propane units of lignin are clearly the limiting factors for the utilization of this material in particleboard.

It thus became clear that a different, but equally or more efficient cross-linking route to be employed in parallel to formaldehyde cross-linking had to be taken if pressing times of use to particleboard mills had to be achieved. Two parallel approaches have been successful. First, methylolated bagasse lignin (MBL) and methylolated kraft lignin (MKL) were reacted in water, with diisocyanate according to a new reaction and mechanism observed for PF resins [22]. Combinations of polymeric MDI, synthetic PF, and methylolated lignins yielded particleboard with full exterior-grade properties at pressing times of as fast as 20 s mm^{-1} when using up to 55% methylolated bagasse lignin [23]. Pressing times using methylolated kraft lignin were also faster but still too slow [23].

Although the reaction of lignins with isocyanates to form urethane in the absence of water is well known (Glasser and Sarkanen, 1989 [24]), the recent findings that raw polymeric MDI does react preferentially with the methylol groups of PF resins in water solution [22,25,26] has opened the possibility to use this rapid, additional cross-linking mechanism to exploit otherwise-slow-reacting phenolic resins, such as methylolated lignin, for adhesives. The possibility of achieving much faster pressing times in adhesive formulations for particleboard based on methylolated lignin has then become a reality. The greatest defect of lignin, too long a pressing time, can then be overcome at least to a great extent.

The same mechanism, proved to work for synthetic phenolic resins, was tried for methylolated kraft and bagasse lignins. In the case of methylolated lignins the mechanism appears to be the same [23] and can be indicated schematically as follows:

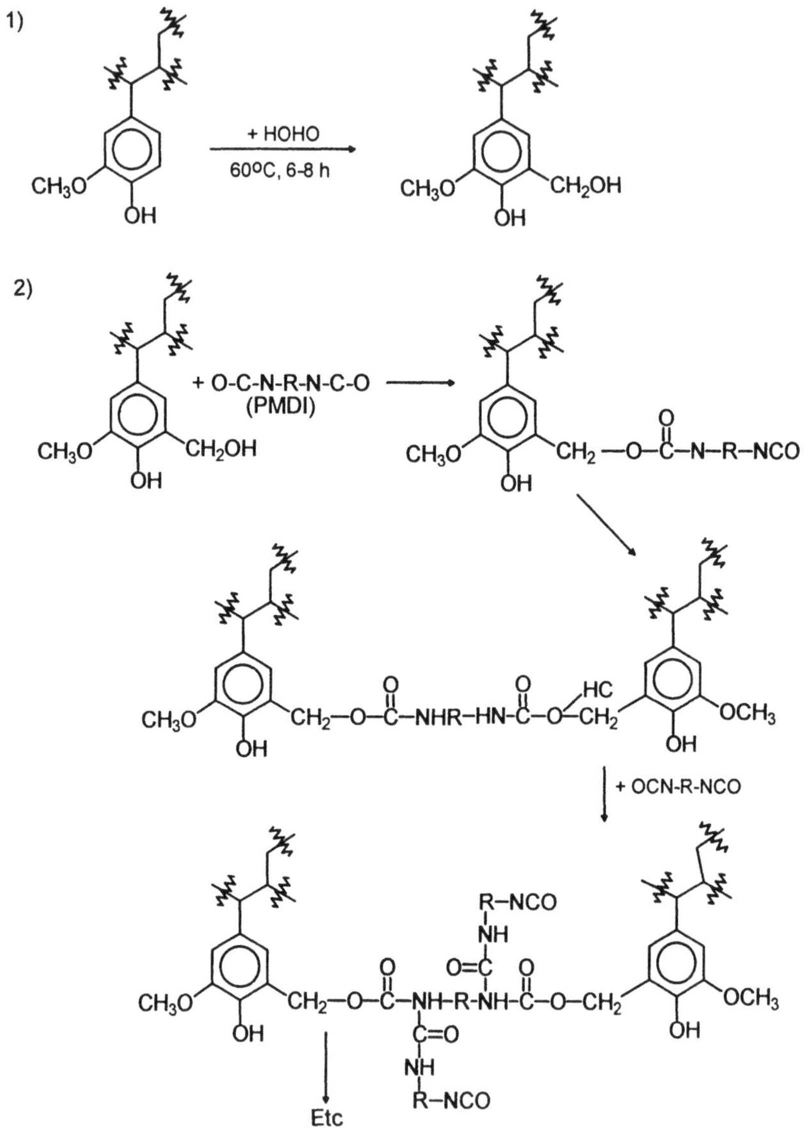

1)

2)

With this schematic representation the greatest number of the cross-links present in the cured resin are urethane bonds, formed by the reaction of isocyanate groups with the methylol groups introduced onto lignin by premethylolation. Additional cross-linking is afforded by isocyanate groups reaction with the –NH– of the urethane bridges already formed [22]. It must be pointed out that in lignins in which all the sites ortho to the aromatic phenolic group are blocked by methoxy groups –OCH_3, reaction of formaldehyde at the meta position would occur. This is a well-known fact observed for the reaction with formaldehyde of any phenols in which all the ortho and para positions are blocked and the meta positions still available [28,29] which was observed initially for mesitol (2,4,6-trimethyl phenol).

Previous work [22,25] on the reaction in water of the methylol group with the isocyanate group has shown that for optimal-strength results a significant number of methylene bridges ($-CH_2-$) or methylene–ether bridges ($-CH_2-O-CH_2-$), obtained by the classical phenolic resin cross-linking mechanism, must also be present in the cured adhesive. The simple reaction of methylolated lignin with polymeric MDI does clearly not lead to any, or to a very low, number of methylene bridges, due to the well-known low reactivity of lignin in formaldehyde-based condensation reactions [2]. However, methylene–ether bridges ($-CH_2-O-CH_2-$) are still very likely to form, as for any phenol with a blocked phenolic hydroxy group [23]. The results obtained might then be far from optimal when using such a simple formulation.

From the above it appears necessary that a method to introduce methylene cross-links, or methylene–ether cross-links, in the system, be it by copolymerization, as in tannin/MDI [25], or by polymer blends, as partially for MDI/PF/tannin [22,23,25], is necessary. The simpler manner to achieve this is to introduce an amount of synthetic PF resin into the MDI/methylolated lignin co-reaction, or alternatively, a TF blend, or even a mixture of TF blend/synthetic PF. If a synthetic PF resin is introduced, both urethane linkages formed between –NCO and –CH_2OH of the synthetic PF, as well as PF to PF and PF to methylolated lignin methylene cross-links, will be formed. The most appropriate approach, however, would be to increase considerably the proportion of $-CH_2-$ cross-links rather than to add a tannin–paraformaldehyde blend or a mixture of the latter blend and of a synthetic PF resin. Two of these systems were tried, resulting in a basic adhesive formulation composed of methylolated lignin (ML): synthetic PF/MDI = 46:24:30 by mass and ML/PF/tannin/MDI = 46:22:12:20 by mass.

In Fig. 6.1 are shown the gel times as a function of pH for the PF resin alone, methylolated kraft lignin (MKL) alone, the PF/MDI 70:30 system, the MKL/MDI 70:30 system, and the MDI/PF/MKL 30:24:26 system. Several interesting features are noticeable: the MDI/PF and MDI/MKL mixes always have much faster gel times than those of PF and MKL resins alone. This is in itself further proof of the existence and effectiveness of cross-linking by urethane bridges even in the presence of excess water

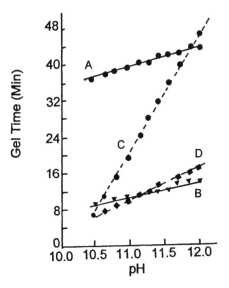

Figure 6.1   Detailed gel time versus pH curves in the pH region 10 to 12 for synthetic PF resol alone (A), MDI/synthetic PF resol 30:70 mix (B), methylolated softwood kraft lignin resol alone (C), and MDI/methylolated softwood kraft lignin resol 30:70 (D). (From Ref. 23.)

[22,23,25,26]. The curves of the lignin resol alone and the MDI/lignin resol system in Fig. 6.1 have very different slopes: the lignin resol curve has a much steeper slope. It is important to note from these curves the hardening acceleration effect imparted by MDI to a lignin resol. Equally significant is the fact that in the greatest part of the pH range examined, MKL presents gel times faster than those of PF resins. This is a clear indication that the poor board results obtained previously with methylolated lignins, both alone and in conjunction with a minority proportion of PF resins, are not alone to blame for slow lignin reactivity with formaldehyde. They should generally be ascribed to the lower number of methylol groups introduced in lignin, to lignin structural characteristics, and hence to poorer cross-links density in the hardened lignin resin. In this respect the sharper slant of the curve of MKL alone in comparison to that of PF alone indicates, on top of what has already been explained before, that a much lower number of methylol groups is likely to be present on methylolated lignin than on the PF resin, a fact that has already been ascertained by formaldehyde absorption measurements.

Second, as a consequence of the elucidation of PF α-set acceleration mechanisms [30], pressing times as short as 7.5 s mm^{-1} for MBL and 10 s mm^{-1} for MKL were obtained, at a lignin content of the resin as high as 65% of total adhesive [31]. These pressing times are faster than for synthetic phenolic resins and of almost the same order of magnitude as for UF resins.

Recent clarification of the mechanism of acceleration of the curing of PF resols under very alkaline reaction conditions by the use of esters [30] provided the means to accelerate further the curing of lignin-based adhesive. The mechanism of acceleration in the case of one of the most suitable esters, propylene carbonate, can be expressed schematically as follows:

with the –CH$_2$OH (methylol) groups still able to react with the –NCO group of MDI to form urethane cross-links at the accelerated rate already described [22,30,31], the diaromatic keto group, in principle still being able to react with a further aromatic ring [30,31]. The acceleration mechanism is then based on a further, third, and faster additional cross-linking reaction and in the consequent apparent increase of functionality of the monomer leading to much earlier, and hence faster, tridimensional cross-linking and gelling [32,33].

The interesting question with lignin is, however, whether such a mechanism should accelerate the curing of a lignin-based resin, particularly for kraft lignins in which the ortho and para sites on the phenylpropane unit aromatic nuclei are already mostly blocked (syringyl moieties). It is already well understood from phenolic resin chemistry that the methylol groups introduced in lignins by their reaction with formaldehyde will occupy the few ortho positions available on the lignin quaiacyl units [2,24,28,34]. Reaction of formaldehyde at the position meta to the phenolic hydroxyl is also well known to occur, with phenols such as mesitol in which all the ortho and para positions are blocked [28,29]. It is, however, much slower and not capable of saturating all available meta positions available on lignin with methylol groups. The few available methylol groups, at whatever site, will react rapidly with the isocyanate group to form urethane bridges and thus proceed on their own reaction route without being involved in the ester acceleration. Thus in kraft lignin, and even in other methylolated lignins, the ester acceleration is likely to involve the free, available metal sites of the aromatic ring, and this has been found to be the case [31].

As already indicated for synthetic PF resins [30], the acceleration effect due to propylene carbonate and other esters is due, also in lignin, to two effects: (1) the presence of a faster, additional, cross-linking reaction in the system, and (2) an effective increase in functionality of the monomer. While a phenyl propane unit with one methylol group on it can be considered, on average, only slightly more than monofunctional or at best difunctional, addition of the ester shifts its functionality to a minimum of over 2 and quite likely of almost 3 [33]. This causes much earlier tridimensional cross-linking and gelling.

The applied results confirm that the ester accelerates considerably the gelling reaction of MKL. Comparable results are obtained with bagasse lignin. Thus in Fig. 6.2 are shown the gel time versus pH, in an alkaline environment, of MKL alone and with various percentages of some ester accelerators. The effect is additional and independent of the diisocyanate acceleration effect, causing additional acceleration of gelling and curing, as can also be seen in Fig. 6.2. The results obtained for exterior particleboard with different proportions of MDI/PF/MKL/triacetin (ester) were excellent and confirmed the theory [31]. An interesting finding from applying to methylolated lignin an α-set ester acceleration mechanism is that if MDI is absent, a 40:60 by mass PF/MKL adhesive still gives excellent exterior-grade particleboard when triacetin ester (glycerol triacetate) is added. However, while press times are much faster than those achievable earlier, such press times are still longer than those obtainable with synthetic PF resins.

There is a last minor point of immediate interest. In particleboard manufacture, triacetin is a better accelerator to use than propylene carbonate, methyl formate, or vinyl acetate [31] because even at higher proportions of triacetin than those reported, a more than adequate pot life of the resin mix at ambient temperature is obtained. Propylene carbonate and worse, methyl formate, although affording the possibility of slightly faster press times, unduly shorten the resin pot life when used in comparable amounts. Results equal to those of triacetin can, of course, also be obtained with propylene carbonate when its percentage on resin solids is substantially decreased, a fact the causes of which have already been explained at length. Adhesives based on methylolated bagasse lignin gave comparable results [31].

While α-set acceleration by esters has been attempted successfully [31], another system of acceleration is also of importance in the field of methylolated lignin adhesives. As MDI/PF/MKL, MDI/TF/MKL, and MDI/MKL

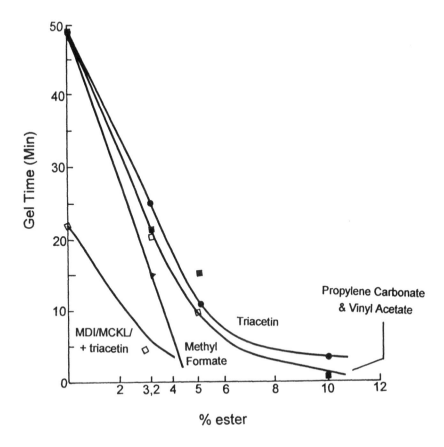

Figure 6.2 Gel times of methylolated kraft lignin (MKL) versus percentage esters at constant pH (pH 10.8 to 11.6).

are all based on reaction of the isocyanate group with the methylol groups of lignin, acceleration by classical urethane-formation catalysts has also been attempted successfully [35–38]. Both a tin catalyst, dibutyltin dilaureate (DBTDL), and an amine catalyst, triethyl amine, were tested with MKL-based formulations. The acceleration effect was very noticeable, as this system of acceleration has an increasingly noticeable effect the slower the phenolic resol, in this case MKL, used [35–38]. It is particularly well suited for methylolated lignins, acceleration of up to 40 times of the reaction having been observed [35–38]. The particleboard results obtained were, however, not as good as expected at pressing times as fast and faster than those obtainable with synthetic phenolic resins alone [35–38]. Of particular interest also is the fact that while these polyurethane synthesis and curing accelerators are normally used in nonaqueous systems, in the case reported they were equally effective in the aqueous systems used [36,38] for particleboard adhesives: a clear case of reaction at the interface. The mechanisms reported for this acceleration also indicate a slight slowing of the reaction with increasing reaction temperatures [37]. The amine catalysts were the most effective [36–38].

V. LIGNIN IN COMBINATION WITH AMINOPLASTIC RESINS

In 1965, W. Arnold [2] found that the pressing time of SSL particleboard obtained with sulfuric acid as catalyst (see the discussion of the Shen procedure in Section III.B) can be reduced by 50% and the specific gravity of the boards by 7 to 10% if 10 to 30% of the SSL blended wood chips are replaced by UF blended wood chips. However, at pressing times of 0.6 to 1 min per millimeter of board thickness, the pressed particleboard has to be posttreated at high temperatures to meet German standards for mechanical strength properties. Again, the necessary posttreatment of the boards have hindered the practical application of this finding. On the other hand, small amounts (up to 10% of the UF resin) of SSL improve the cold adherence and tack of a blended UF particleboard (Schmidt-Hellerau, 1973, 1977) [2]. This has found practical application and is in current industrial use in some western European particleboard mills.

Roffael [7] has shown that 20% of the UF in the surface layers of UF particleboard can be replaced by ammonium-based lignosulfonate without significantly worsening the mechanical board properties. The release of formaldehyde decreased only slightly, which was attributed to the reduction of UF resin rather than to a reaction of formaldehyde with the lignin. The binding of formaldehyde by lignin in UF particleboard is claimed in three Japanese patents [2], together with other patents dealing with lignin UF formulations as wood adhesives. Other improvements achieved by ligno-sulfonate in UF resins are decrease in adhesive viscosity, wettability of wood particles, and water resistance of finished boards.

According to one report [2], substitution by SSL of up to 15% of the UF adhesive in particleboard does not cause any major impairment in particle-board properties. This can be seen from the properties of 17.7-mm-thick one-layer particleboard obtained with 8% UF binder (F/U = 1.27), replaced partially by 10 to 30% magnesium-based SSL, as listed. The adhesive contained 0.5% paraffin emulsion and 3% ammonium chloride. The pressing time was 10 s mm^{-1} at 200°C [2]. It is, however, obvious that substitution of 20% or more of the UF binder by magnesium-based SSL worsens both the strength and water resistance of the boards, while the gelling time (pot life) of the adhesive is increased. When three-layer 20-mm particleboard was manufactured with 15% of calcium-based (A), sodium-based (B), or ammonium-based (C and D) lignosulfonate and 85% UF binder, with pressing times of 9 s mm^{-1} at 200°C and 10.5% adhesive in the surface layer and 8.5% in the core layer, the board properties were different.

At board densities of about 0.7 g cm^{-3}, the bending and tensile strengths of the UF–LS boards are not decreased compared to boards prepared with 100% UF binder, while the percentage of swelling is increased. The formaldehyde release is considerably decreased by ammonium (C and D)-rather than by calcium- and sodium-based SSLs, indicating that the ammonium ions, but not the lignin, react with formaldehyde under the conditions existing in the boards. In the case of calcium- and sodium-based SSLs, the reduction in formaldehyde release lies between 10 and 18%, which corresponds to the amount of SSL in the UF–SSL formulation.

In a recent patent by Edler (1978) [2] it has been claimed that about 33% of UF binder in particleboards can be replaced by ammonium-based SSL if certain conditions are maintained. First, the UF resin should have a relatively high number of methylol (CH_2OH) groups, characterized by a Witte number of 1 to 1.8, preferably 1.6, which leads to better compatibility between UF and SSL. Second, the concentration of ammonium ions has to be adjusted to 0.2 to 4%. The ammonium ions react with free formaldehyde, forming less reactive hexamethylenetetramine, which leads to excessive sulfonic acid groups in lignin. If the ammonium ion concentration is higher than 4%, based on dry lignosulfonate, the acidity becomes too high, resulting in very fast curing. The latter causes soft board surfaces and diminished strength properties. On the other hand, if the ammonium ion concentration is below 0.2%, the curing time becomes too long. The properties of four types of particleboard obtained with three different types of adhesive were recorded: A having a Witte number of 1.58, B of 1.50, and C of 1.02, the latter prepared using only UF resin, while types A and B contained 33% lignosulfonate and 67% UF. Pressing times were about 15 s mm^{-1} at 160°C. Wood chips consisting of 67% pine and 33% Douglas fir gave different results. The mechanical strength properties of boards obtained with resins types A and B (33% lignosulfonate) show no major impairment compared with those of conventional UF boards (resin type C). Significant improvements in water resistance are gained, due to the polyphenolic structure of lignin [2].

An interesting approach to melamine–lignin adhesives has been advocated. The methyl ethers of melamine (i.e., hexamethoxymethyl melamine) can also copolymerize with hydroxyl-containing molecules [39,40]. Thus it can be used as a cross-linking and fortifying system for lignin-based wood adhesives, by reaction with both hydroxyl groups of lignin and hydroxybenzyl alcohol groups of methylolated lignins [39,40]. However, this approach has the disadvantage that a reaction occurs between the methyl ethers of melamine and a hydroxyl-carrying compound and is favored in nonaqueous systems, which is a definite drawback [41].

VI. ^{13}C NMR OF METHYLOLATED LIGNINS

^{13}C NMRs of the methylolation of both bagasse and softwood kraft lignin show clearly that during the reaction with formaldehyde, methylol groups are formed. Thus, from Figs. 6.3 and 6.4 the three bands at 120 ppm (ArC–H) decrease drastically and almost disappear, indicating extensive introduction of substituted groups on the aromatic ring of the lignin phenylpropane unit. This is substantiated by the increase of the bands at around 130 ppm, indicating that ArC–C linkages form from the methylene and methylene ether bridges and by the strong appearance of the bands at 51 ppm indicating –CH$_2$–. The

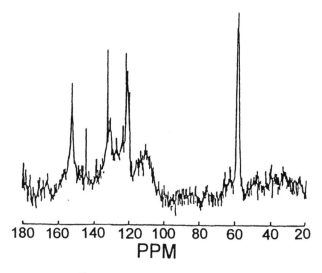

Figure 6.3 ^{13}C NMR spectrum of bagasse lignin.

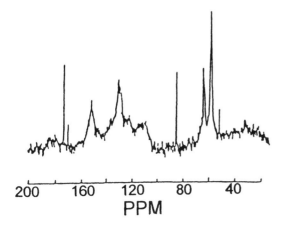

Figure 6.4 ^{13}C NMR spectrum of methylolated bagasse lignin resol.

strong increase in the peak at 64 to 65 ppm indicates formation of a considerable number of –CH$_2$OH groups on the aromatic ring. The broad, strong band at 58 to 59 ppm increases in relative intensity, indicating that on the original –OCH$_3$ groups is superimposed a band indicating formation of –CH$_2$–O–CH$_2$– bridges. There is a sharp peak at 84.8 ppm denoting the

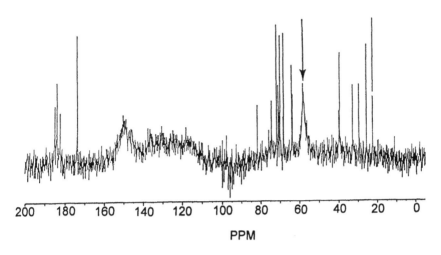

Figure 6.5 ^{13}C NMR spectrum of kraft lignin in D$_2$O.

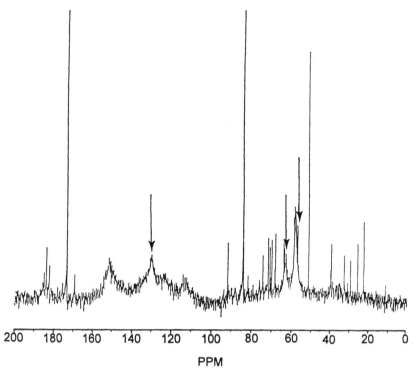

Figure 6.6 ^{13}C NMR spectrum of methylolated kraft lignin in D_2O. Note the free formaldehyde peak at 84.5 ppm, the strong appearance of a new broad peak at 130.4 ppm, the appearance of peaks at 63.2 and 58.7 ppm, and the general strong increase in the series of bands at 63.9 to 64.5 ppm relative to the preexisting pattern of the spectrum.

presence of free HCHO, although this does not appear to be excessive, and one peak at 174 ppm denoting formation of formic acid by Cannizzaro reaction. The trends shown in Figs. 6.3 and 6.4 are equally present in the ^{13}C NMR of methylolated softwood kraft lignin.

^{13}C NMR spectra also indicate that both formaldehyde and ester α-set attacks can occur at the 3- and 5-positions of the aromatic nuclei of the phenyl propane units in methylolated kraft and bagasse lignins. ^{13}C NMR spectra of kraft lignin and methylolated kraft lignin indicate that both ortho and meta methylolations have occurred (Figs. 6.5 and 6.6). Thus the corresponding sizable apparent increase in the 58.73-ppm band of the methylene carbon of

the methylol group in meta methylolation and the appearance of bands at 63.2 and 63.9 ppm for ortho methylolation indicate that both types of attack have occurred. The existence of a considerable number of methylol groups at the C5 position of the phenyl propane unit is confirmed by the appearance of the very strong series of bands at 130.44 ppm, which is clearly visible when comparing Figs. 6.5 and 6.6.

REFERENCES

1. F. Melms and K. Schwenzon, *Verwertungsgebiete der Sulfitablauge*, VEB Deutscher Verlag für Grunstoffindustrie, Leipzig, 1967.
2. H. H. Nimz, Chapter 5 in *Wood Adhesives: Chemistry and Technology*, Vol. 1, (A. Pizzi, ed. Marcel Dekker, New York, 1983.
3. K. Freundenberg, *Science*, *148*: 595 (1965); K. Freundenberg, in *Constitution and Biosynthesis of Lignin* (K. Freudenberg and A. C. Neish, eds.), Springer-Verlag, Berlin, 1968.
4. K. C. Shen and L. Calvé, *Adhes. Age*, 25 (Aug. 1980).
5. R. J. Mahoney, *Proceedings of the 14th International Particleboard Symposium*, Washington State University, Pullman, Wash., 1980.
6. K. G. Forss and A. Fuhrmann, *Paperi ja Puu (Finland)*, *11*: 817 (1976).
7. E. Roffael, *Adhaesion*, *11*: 334; *12*: 368 (1979).
8. E. Roffael and W. Rauch, 1974, *Holz Zentralbl.*, No. 43/44, Apr. 10, 1974.
9. D. P. C. Fung, K. C. Shen, and L. Calvé, *Spent Sulphite Liquor–Sulphuric Acid Binder: Its Preparation and Some Chemical Properties*, Report OPX 180 E, Eastern Forest Products Laboratory, Ottawa, 1977.
10. K. C. Shen and M. N. Carroll, *For. Prod. J.*, *19*(8): 17 (1969); K. C. Shen, *For. Prod. J.*, *24*(2): 38 (1974).
11. *Canadian Standard for Particleboard*, CSA 0188/68.
12. K. C. Shen, *For. Prod. J.*, *27*(5): 32 (1977).
13. H. H. Nimz and G. Hitze, *Cellul. Chem. Technol.*, *14*: 371 (1980).
14. E. Roffael and W. Rauch, *Holzforschung*, 26: 197 (1972).
15. E. Roffael and W. Rauch, *Holzforschung*, 25: 149 (1971).
16. E. Roffael and W. Rauch, *Holzforschung*, 27: 214 (1973).
17. K. G. Forss and A. Fuhrmann, *For. Prod. J.*, *29*(7): 39 (1979).
18. *LCHW-Ligninsulfonate in Holzspanplatten*, Studie 2 and 3, Lignin-Chemie Waldhof-Holmen BmbH, Dusseldorf, Germany, 1979.
19. P. Md. Tahir and T. Sellers, Jr., *19th IUFRO World Congress*, Monreal, Canada, Aug. 1990.
20. L. Calvé, *19th IUFRO World Congress*, Monreal, Canada, Aug. 1990.
21. A. Pizzi, F. A. Cameron, and G. H. van der Klashorst, Chapter 7 in *Adhesives from Renewable Resources* (R. W. Hemingway, A. Conner, and S. J. Branham, eds.), ACS Symposium Series No. 385, American Chemical Society, Washington, D.C., 1988.
22. A. Pizzi and T. Walton, *Holzforschung*, *46*(6): 541 (1993).

23. A. Stephanou and A. Pizzi, *Holzforschung*, *47*(5): 439 (1993).
24. W. G. Glasser and S. Sarkanen, *Lignin: Properties and Materials*, ACS Symposium Series No. 397, American Chemical Society, Washington, D.C., 1989.
25. A. Pizzi, J. Valenzuela, and C. Westermeyer, *Holzforschung*, *47*(1): 68 (1993).
26. A. Pizzi, E. P. von Leyser, J. Valenzuela, and J. G. Clark, *Holzforschung*, *47*(2): 168 (1993).
27. K. G. Frisch, L. P. Rumao, and A. Pizzi, in *Wood Adhesives: Chemistry and Technology*, Vol. 1 (A. Pizzi, ed.), Marcel Dekker, New York, 1983.
28. N. J. L. Megson, *Phenolic Resin Chemistry*, Butterworth, London, 1958.
29. S. R. Finn and J. W. G. Musty, *J. Chem. Ind. London*, *69*, Suppl. issue 1: S3 (1950).
30. A. Pizzi and A. Stephanou, *J. Appl. Polym. Sci.*, *49*: 2157 (1993).
31. A. Stephanou and A. Pizzi, Part II, *Holzforschung*, *47*(6): 501 (1993).
32. P. J. Flory, *Principles of Polymer Chemistry*, Cornell University Press, Ithaca, N.Y., 1953.
33. R. B. Seymour and C. E. Carraher, *Polymer Chemistry*, Marcel Dekker, New York, 1988.
34. A. Pizzi, G. H. van der Klashorst, and F. A. Cameron, Chapter 7 in *Adhesives from Renewable Resources (R. W. Hemingway and A. H. Conner, eds.)*, ACS Symposium Series No. 385, American Chemical Society, Washington, D.C., 1989.
35. R. Krause, B.Sc. (Hons.) research report, University of the Witwatersrand, Johannesburg, South Africa, 1993.
36. P. Cheesman, B.Sc. (Hons.) report, University of the Witwatersrand, Johannesburg, South Africa, 1993.
37. D. B. Batubenga, M.Sc. thesis, University of the Witwatersrand, Johannesburg, South Africa, 1994.
38. D. B. Batubenga, A. Pizzi, A. Stephanou, P. Cheesman, and R. Krause, *Holzforschung*, in press (1994).
39. S. S. Kelley, P. C. Muller, W. H. Newman, and W. G. Glasser, in *Wood Adhesives in 1985: Status and Needs*, Forest Products Research Society, Madison, Wis., 1986.
40. W. J. Blank, *J. Coating Technol.*, *51*(656): 61 (1979).
41. E. S. Ntsihlele, M.Sc. thesis, University of the Witwatersrand, Johannesburg, South Africa, 1993.

7

Resorcinol Adhesives

I. INTRODUCTION

Resorcinol–formaldehyde (RF) and phenol–resorcinol–formaldehyde (PRF) cold-setting adhesives are used primarily in the manufacture of structural glulam, finger joints, and other exterior timber structures. They produce bonds not only of high strength but also of outstanding water and weather resistance when exposed to many climatic conditions [1,2]. PRF resins are prepared primarily by grafting resorcinol onto the active methylol groups of the low-condensation resols obtained by the reaction of phenol with formaldehyde. Resorcinol is the chemical species that gives to these adhesives their characteristic cold-setting behavior. It gives accelerated and improved cross-linking, at ambient temperature and on addition of a hardener, not only to the resorcinol–formaldehyde resin but to the PF resins onto which resorcinol has been grafted by chemical reaction during resin manufacture. Resorcinol is an expensive chemical, produced in very few locations around the world (to date, only three commercial plants are known to be operative: in the United States, Germany, and Japan), and its high price is the determining factor in the cost of RF and PRF adhesives. It is for this reason that the history of RF and PRF resins is closely interwoven, by necessity, with the search for a decrease in their resorcinol content, without loss of adhesive performance.

In past decades, significant reductions in resorcinol content have been achieved: from pure RF resins, to PRF resins in which phenol and resorcinol were used in equal or comparable amounts, to the modern-day commercial resins for glulam and finger jointing in which the percentage, by mass, of resorcinol on liquid resin is on the order of 16 to 18%. A step forward has been the development and commercialization of the "honeymoon" fast-set system [3], coupled with the use of tannin extracts which in certain countries are used to obtain PRFs of 8 to 9% resorcinol content without loss of performance and with some other advantages. This was a "system" improvement, not an advance on the basic formulation of PRF resins.

A. Chemistry and Application of RF Resins

The same chemical mechanisms and driving forces presented for phenol–formaldehyde resins apply to resorcinol resins. Resorcinol reacts readily with formaldehyde to produce resins, which harden at ambient temperatures if formaldehyde is added. The initial condensation reaction, in which A-stage liquid resins are formed, leads to the formation of only linear condensates when the resorcinol/formaldehyde molar ratio is approximately 1:1 [4]. This reflects the reactivity of the two main reactive sites (4- and 6-positions) of resorcinol [5]. However, reaction with the remaining reactive but sterically hindered site (2-position) between the hydroxyl functions also takes place [4]. In relation to the weights of RF condensates which are isolated and on a molar basis, the proportion of 4- plus 6-linkages relative to 2-linkages is 10.5:1. However, cognizance must be taken of the fact that the first-mentioned pair represents two condensation sites relative to one. The difference in reactivity of the two types of sites (i.e., 4- or 6-position relative to the 2-position) is then 5:1 [4]. Linear components always appear to form in preference to branched components in A-stage resins [4]; that is, terminal attack leads to the preferential formation of linear rather than branched condensates. This fact can be attributed to:

1. The presence of two reactive nucleophilic centers on the terminal units as opposed to single centers of doubly bound units already in the chain.
2. The greater steric hindrance of the available nucleophilic center (nearly always at the 2-position) of the doubly bound units as opposed to the lower steric hindrance of at least one of the nucleophilic centers of the terminal units (a 4- or 6-position is always available). The former is less reactive as a result of the increased steric hindrance. The latter are more reactive.
3. The lower mobility of doubly bound units, which further limits their availability for reaction.

most common type of resorcinol /
formaldehyde "tetramer"

The absence of methylol ($-CH_2OH$) groups in all six lower-molecular-weight compounds isolated [4] reflects the high reactivity of resorcinol under acid or alkaline conditions. It also shows the instability of its *p*-hydroxybenzyl alcohol groups and its rapid conversion to *p*-hydroxybenzyl carbonium ions or quinone methides. This explains how identical condensation products are obtained under acid or alkaline reaction conditions [4]. In acid reaction conditions methylene-ether-linked condensates are also formed, but they are highly unstable and decompose to form stable methylene links in periods of between 0.25 and 1 h at ambient temperature [6,7].

From a kinetic point of view, the initial reaction of condensation to form "dimers" is much faster than the later condensation of these dimers and higher polymers. The reaction of resorcinol with formaldehyde on an equal molar basis and under identical conditions proceeds at a rate that is approximately 10 to 15 times faster than that of the equivalent PF system [6,13]. The high reactivity of the RF system renders it impossible to have these adhesives in resol form. Therefore, only RF novolaks, resins not containing methylol groups, can be produced. Thus all the resorcinol nuclei are linked together through methylene bridges with no methylol groups or methylene–ether bridges.

The reaction rate of resorcinol with formaldehyde is dependent on the molar ratio of the two constituents, the concentration of the solution, the pH, the temperature, the presence of various catalysts, and the amount of certain types of alcohols present [8–11]. The effect of pH and temperature on the reactivity of the RF system is shown in Fig. 7.1 [11,12]. Methanol and ethanol slow the rate of reaction. Other alcohols behave similarly, the extent of their effect being dependent on their structure. Methanol lengthens gel time more than that of other alcohols; higher alcohols are less effective. The retarding effect on the reaction is due to temporary formation of hemiformals between the alcohols and the formaldehyde. This reduces the reaction rate because of the

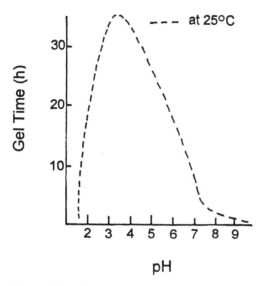

Figure 7.1   Typical gel time as a function of pH of resorcinol for RF adhesives.

lower concentration of available formaldehyde [9,10]. Other solvents also affect the rate of reaction by forming complexes, or by hydrogen bonding with the resorcinol [9,12].

In the manufacture of pure resorcinolic resins, the reaction would be violently exothermic unless controlled by the addition of alcohols. Because the alcohols perform other useful functions in the glue mix, they are left in the liquid glue. PRF adhesives are generally prepared by reaction of phenol with formaldehyde to form a polymer that has been proved to be in the greatest percentage and often completely linear [4]. This can be represented as follows:

$$I$$

m > 0 in integer numbers

o ◄n► 2 in integer numbers

In the reaction, the resorcinol chemical is added in excess, in a suitable manner, to polymer I to react with the –CH$_2$OH groups to form polymers of the following type, in which the terminal resorcinol groups can be resorcinol chemical or any type of RF polymer.

Where straight resorcinol adhesives are not suitable, resins can be prepared from modified resorcinol [12]. Characteristic of these types of resins are those used for tire cord adhesives, in which a pure RF resin is used, or alternatively, alkylresorcinol or oil-soluble resins suitable for rubber compounding obtained by prereaction of resorcinol with fatty acids in the presence of sulfuric acid at high temperature followed by reaction with formaldehyde. Worldwide, more than 90% of resorcinol adhesives are used as cold-setting wood adhesives; of the other applications, tire cord adhesive is most notable, and constituting no more than 5% of the total use.

Together with the more traditional finger-jointing adhesives [14], a series of ambient-temperature fast-setting separate application systems have been developed. These eliminate the long delays caused by the use of more conventional PRF adhesives which require lengthy periods to set. These types of phenolic adhesives are applied separately. They were first developed in the United States [15–18] to glue large components where presses were impractical. Kreibich [18] describes these separate applications or "honeymoon" systems as follows: "Component A is a slow-reacting PRF resin with a reactive hardener. Component B is a fast-reacting resin based on *m*-aminophenol with a slow-reacting hardener. When A and B are mated, the reactive parts of the components react within minutes to form a joint which can be handled and processed further." Full curing of the slow-reacting part of the system takes place with time. *m*-Aminophenol is a frightfully expensive chemical, and for this reason, these systems were discarded and not used industrially [14]. In their original concept, component A is a traditional PRF cold-setting adhesive at its standard pH of 8 to 8.5 to which paraformaldehyde hardener has been added. Flour fillers may be added or omitted from the glue mix. Component B is a phenol/*m*-aminophenol/formaldehyde resin with a very high pH (and therefore a high reactivity), which contains no hardener or only a very slow hardener.

More recently, a modification of the system described by Kreibich has been used extensively in industry with good success. Part A of the adhesive is again a PRF cold-setting adhesive, with powder hardener added at its standard pH. Part B can be either the same PRF adhesive with no hardener and the pH adjusted to 11, or a 50–55% tannin extract solution at a pH of 11–12, provided that the tannin is of the condensed or flavonoid type, such as wattle, quebracho, hemlock, or pine extract, with no hardener [3,19]. The results obtained with these two systems are good and the resin has not only all the advantages desired but is also considerably cheaper as a result of the use of vegetable tannins and of the halving of the resorcinol content of the entire adhesive system [3,19,20].

II. BRANCHED LOW-RESORCINOL-CONTENT PRFs [21–23]

Traditional PRF adhesives are generally prepared by reaction of phenol with formaldehyde to form PF resols, which are mostly linear. Resorcinol is added to this resin often in excess of the $-CH_2OH$ methylol groups of the PF resin. Resorcinol is then linked by methylene bridges ($-CH_2-$) to the PF resin skeleton. Thus in the PRF resin the linear PF resin oligomers are in general terminated by resorcinol moieties. The PRF resin produced can be represented schematically as

resorcinol–CH_2–[phenol–$CH_2]_n$–resorcinol

If a chemical molecule capable of branching (three or more effective sites with an aldehyde) is used during or after the preparation of the PF resin, a branched PF, and consequently a branched PRF resin, is produced. This can be represented schematically as

resorcinol–[–CH_2–phenol–]$_n$–CH_2 $\quad$ CH_2–[–phenol–CH_2–]$_n$–resorcinol

$$\diagdown \quad \diagup$$

branching
molecule

$|$

CH_2–[–phenol–CH_2–]$_n$–resorcinol

with $n \geq 1$ and an integer number and comparable to, similar to, or equal to n in the preceding scheme for the production of PRF resins.

It is notable when comparing linear and branched resins that for every n molecules of phenol used in the particular schematic examples shown, two molecules of resorcinol are used in the case of a normal, traditional, linear PRF adhesive, whereas only one molecule of resorcinol for n molecules of phenol is used in the case of a "branched" PRF adhesive. The amount of resorcinol has then been halved or approximately halved in the case of the branched PRF resin. A second effect caused by the branching is the noticeable increase in the degree of polymerization of the resin. This causes a considerable increase in the viscosity of the liquid adhesive solution. Because PRF adhesives must be used within fairly narrow viscosity limits, to return to a liquid PRF adhesive viscosity level within these limits, the resin solids content in the adhesive must be lowered considerably, with a consequent further decrease in total liquid resin of the amount of resorcinol and of the other materials, except solvents and water. This further decreases the cost of the resin without decreasing its performance.

To conclude, the decrease in resorcinol by branching of the resin is based on two effects:

1.   The decrease of resorcinol percentage in the polymer itself, hence on the resin solids, due to the decrease in the number of PF terminals onto which resorcinol is grafted during PRF manufacture.
2.   The increase in molecular weight of the resin, which, by the need to decrease the percentage resin solids content to a workable viscosity, decreases the percentage of resorcinol in the liquid resin (not on resin solids).

It is clear that in a certain sense a branched PRF will behave as a more advanced, almost precured, phenolic resin. While the first effect described is a definite advance on the road to better engineered PRF resins, the second effect can be obtained with more advanced (reactionwise) linear resins. The contribution of the second effect to the decrease in resorcinol is not less marked than that of the first effect. It is, however, the second effect that accounts for the differences in behavior between branched and linear PRF adhesives. Branching molecules could be resorcinol, melamine, urea, and others [21]. Urea is favored because it is much cheaper than the others and needs to be added only in 1.5 to 2% of total resin. When urea is used as a brancher, the adhesive assumes after a few days an intense and unusual (for resorcinol resins) blue color—hence its nickname, "blue glue."

Later work [22,23] has shown, however, that tridimensional branching has very little to do with the improved performance of these low-resorcinol-content adhesives, with tridimensionally branched molecules contributing, in the best cases, not more than 8 to 9% of total strength [22,23]. In reality, the addition of urea causes the reaction foreseen, not in three branching points but only in two sites of the branching molecule. This is equivalent to saying that most of the resin doubles linearly in molecular weight and degree of polymerization, while the final effect, good performance at half the resin resorcinol content, is maintained [22,23]. This effect is based on the relative reactivity for phenolic methylols of urea and of unreacted phenol sites, and thus while the macro effect is as desired, at the molecular level it is only a kinetic effect, due to the various relative reactivities of urea and phenol under the reaction conditions used: thus

$$resorcinol-CH_2-[-phenol-CH_2-]_n-resorcinol$$
$$resorcinol-CH_2-[-phenol-CH_2-]_n-resorcinol$$
$$\downarrow$$

(with urea) resorcinol–CH$_2$–[–phenol–CH$_2$–]$_n$–urea–[–CH$_2$–phenol–]–CH$_2$–resorcinol

Thus the halving of resorcinol content is still obtained, but between 90 and 100% of the polymers in the resin are still linear.

It is noticeable that the same degree of polymerization and doubling effect cannot be obtained by lengthening by any period the reaction time of a PF resin without urea addition [22,23]. These liquid resins then work at a resorcinol content of only 9 to 11%, thus considerably lower than traditional PRF resins. These resins can also be used with good results for honeymoon fast-setting adhesives in PRF–tannin systems, thus further decreasing the total content of resorcinol in the total resin system to a level as low as 5 to 6%.

III. URF ADHESIVES [23,24]

Urea–resorcinol–formaldehyde (URF) cold-setting adhesives of performance comparable with PRF are referred to in the literature [14,25]. These resins rely on the fact that once the adhesive is hardened, the resorcinol network, impervious to water, will protect the aminoplastic bonds of the UF resin from water attack. Recently [23,24], the same branching or rapid molecular doubling applied to PRF resins has also been applied to prepare URF resins of reduced resorcinol content. This is a more difficult undertaking than for PRF resins, because of an additional effect militating against such resins: whereas, in principle, a decrease of resorcinol appears possible (as for the PRF), too low a quantity of resorcinol, while possibly still affording good dry strength of the bonded joint, may not be able to protect the UF backbone of the resin from water degeneration, possibly causing catastrophic wet strength loss strength of joints.

Suppose that the average degree of polymerization of the UF resin before resorcinol addition is n; then the URF resin produced can be represented schematically as [14,24].

resorcinol–CH_2–(–urea–CH_2–)$_n$–resorcinol

where $n \geq 1$ in integer numbers.

If a chemical molecule capable of extensively branching or of rapid doubling linearly of the molecular weight of the UF and URF resins is used after preparation of the UF resin, an URF of lower resorcinol content will be produced. Thus

resorcinol–(–CH_2–urea–)$_n$–CH_2–branching unit –CH_2–(–urea–CH_2–)$_n$–resorcinol
+ small amounts tridimensionally branched on the branching unit

where $n \geq 1$ in integer numbers. The lower resorcinol requirement is achieved by halving the terminal resorcinol needed in the resin as well as by an increase in resin viscosity of the same solids content. One major difference expected from PRF adhesives, which are mostly linear, is that in URF resins tridimensional branching is already present in the basic UF resin, urea being effectively trifunctional in its reaction with formaldehyde. That UF resins are always tridimensionally branched can easily be seen by ^{13}C NMR analysis [26,27]. Thus

$$CH_2\text{–resorcinol}$$
$$|$$
resorcinol–(–CH_2–urea–)$_n$–CH_2–resorcinol

where $n \geq 1$ in integer numbers. If additional branching is introduced by a further branching molecule, the extensively branched URF can be represented as follows:

$$CH_2\text{–resorcinol} \qquad CH_2\text{–resorcinol}$$
$$| \qquad\qquad\qquad |$$
resorcinol–(–CH_2–urea–)$_n$–CH_2 $\quad CH_2$–(–urea–CH_2–)–resorcinol
$$\backslash \quad /$$
$$\text{branching}$$
$$\text{molecule}$$

where $n \geq 1$ in integer numbers.

Thus it is clearly noticeable when comparing normal UFs and additionally branched UF resin that for n molecules of urea used, two or three molecules of resorcinol are used in the first case, whereas only one or two molecules of resorcinol for n molecules of urea are used in the second case. It has already been shown [22] that branched PRF adhesives of low resorcinol content, still presenting good performance, can be prepared. For these the lowest percentage of resorcinol in the liquid resin which still appeared to give a resin consistently satisfying the requirements of international specifications [28,29] for close-contact, cold-setting adhesives for wood with still acceptable pot life, was 10.6% [22,24]. This is significantly lower than the 16 to 18% of today's commercial PRF resins. Many molecules can be used as additional branching units [14]. In the URF resins the additional tridimensional branching units used were resorcinol and a polymeric natural resorcinolic material, a polyflavonoid tannin [24,30], the latter most often being a pentafunctional unit.

Traditional URF resins of acceptable performance operate at 55 to 58% resin solids content. This is slightly higher than for linear PRF resins. The viscosity of these resins at this percentage solids content is low and they cannot be diluted further without deterioration in their performance. Resorcinol-branched URF adhesives operate in a much wider percentage solids content range, such as from 48.8 to 59% resin solids content [24]. In general, resorcinol-induced branching in URF resins does not appear to have a significant effect on the viscosity of the resins. The viscosity of such resins is comparable to that of linear URF resins. Mimosa tannin–branched URF resins are, however, slightly more viscous than the corresponding resorcinol-branched URF resins. This increase in viscosity could be attributed to the higher molecular weight of the wattle tannin macromolecules that have condensed onto the URF resin. These resins also operate in a much narrower range of percentage resin solids content (56 to 58%).

The lowest percentage resorcinol by mass on total liquid resin which gave acceptable performance in traditional, resorcinol-branched, and tannin-branched URF resins in 16, 13.4, and 12.7%, respectively. The results also showed that there is an optimum level of resorcinol for each type of URF resin [24]. URF resins to which resorcinol was added as an additional branching unit, and URFs without an additional branching unit had pot lives longer than that of urea-branched PRF resins [23,24]. URF resins in which mimosa polyflavonoid tannin was the additional branching unit gave, instead, much shorter pot life. Such URF resins presented the same sensitivity to pH decrease as that of their PRF counterparts [23,24]. Again, only acetylsalicylic acid [23,24] was capable of correcting the resin pH downward, with consequent lengthening of pot life. Any other acid, in solution or solid, caused localized precipitation of the resin.

Another problem encountered in URF resins was their stability (shelf life). Whereas traditional and branched PRF resins are stable and display long shelf lives, this was a problem in the production of URF resins. A decrease in U/F molar ratio resulted in a more stable resin with a much longer shelf life. The most stable U/F ratio was found to be 1:0.5. As urea has three reactive sites, and also as a result of its reactivity, most UF resins appear to be naturally branched or potentially branched. This is quite different from traditional PF resins, which appear to be mostly linear [31]. It must be stressed that what are referred to as traditional URF resins are not necessarily linear; they are merely URF resins in which additional induced branching has not been introduced by the addition of another branching unit during its preparation.

On this basis, URF resins to which resorcinol has not been added as a branching unit show an increase in branching on the UF backbone of the resin when total terminal resorcinol is increased. Conversely, URFs in which additional branching has been induced by introduction in midreaction of a small number of resorcinol branching units show a decrease in branching with resorcinol decrease. By substituting the resorcinol brancher with mimosa tannin extract, it was hoped that the percentage resorcinol by mass on total liquid resin would decrease. It was, however, feared that the percentage wood failure results would be lower because condensation of the macromolecule on the URF resin would imply an increase in the molecular weight of the resin.

The applied results showed that if the amount of resorcinol brancher did not prove to be excessive in the branched URF resin, optimum results were also achieved when an equivalent amount of tannin extract was used as a brancher. The surprising result was that the introduction of the tannin extract as a brancher decreased the percentage of resorcinol by mass on total liquid resin without deteriorating the percent wood failure results. A possible explanation of this is the difference in the molecular geometry created when branching occurs on a small resorcinol molecule compared with that created when it occurs on a large tannin macromolecule. When resorcinol is used as a branching molecule it has a much more rigid structure than when tannin extract is used. The molecular geometry of branching in the latter is not as densely packed. The reactive sites on the wattle tannin macromolecule are also more widely spaced than those on resorcinol. The more widely spaced and more flexible molecular geometry introduced when polyflavonoid tannin is used as a brancher may be allowing better adhesion of the resin to the wood, resulting in higher percentages of wood failure. Ordinarily, higher molecular weights cause poorer wetting, hence poorer wood failure, but in this case it appears that the molecular geometry caused by branching molecules has a significant effect on the ability of the adhesive to wet the substrate.

The amount of methanol used in the solvent also affected the performance of both traditional and branched URF resins. In traditional URF resins it was observed that if the amount of methanol in the solvent used was increased, the percentage wood failure in the dry test improved, but the water-resistant properties of the resin deteriorated. Methanol is known to be a retarder in UF resin synthesis. Methanol retards the process whereby paraformaldehyde releases formaldehyde, which can then react with urea to form methylol groups. As fewer reactive methylol groups result, shorter UF polymer chains are formed. Specific adhesion to lignocellulosic substrates of lower-molecular-weight resins is greater [32]. For this reason it was believed that higher percentages of wood failure in the dry test would result, as indeed they did. However, it has been reported [33] that methanol content makes a considerable difference in the water resistance of UF resins. This effect is due to the formation of methylated UF resins, which on curing remain unchanged and because of their easy solubility in water, destroy the water resistance of the cured, bonded joint. The same effect is observed in traditional URF resins in which 45% by mass of methanol was used in the solvent mixture [23,24].

In conclusion, effects that appear to influence the results of branched URF resins include [23,24]:

1. Amount of brancher.
2. Type of brancher, with polyflavonoid tannins allowing greater decrease of resorcinol content.
3. Temperature of reaction, with lower reaction temperatures (50 to 60°C) giving much better results.
4. Amount of final resorcinol, with resorcinol contents as low as 12 to 13% still giving resins of acceptable performance.
5. Amount of methanol used in solvent, with increasing methanol increasing the percentage of wood failure but progressively impairing the water resistance of the cured linear resins but not of the cured branched resins.
6. Physical properties such as pH and viscosity.

Although these effects also influenced the results of branched PRF resins [22,23], it appears that the range in which these effects overlap to produce acceptable results in branched URF resins is much narrower than in branched PRF resins.

IV. TRF RESINS OF REDUCED RESORCINOL CONTENT

Approaches other than the honeymoon fast-sets and the branching adopted for synthetic PRFs and URFs have also been found to reduce resorcinol content. An unusual and interesting approach is one used for tannin–resorcinol–formaldehyde (TRF) cold-sets which relies on the liberation of the resorcinol from the structure of the tannin itself to yield a decrease in the resorcinol chemical that needs to be added to the final resin. Recently, rearrangements of flavonoid tannins which led to the formation of phlobatannins have been described in the literature [37]. Reactive resorcinol still linked to the framework of the rearranged flavonoids is generated by an acid treatment, such as in the formation of phlobatannins or "tanner's red," well known in tanning practice. This rearrangement is also achieved by milder routes, such as treatment of the tannin with $NaHCO_3$–Na_2CO_3 buffer of 50°C and pH 10 for a few hours [37]. From this, several options were identified to prepare mimosa tannin cold sets through the phlobatannin route [38,39]. Of the several routes tried [38,39], treatment after preparation of the TRF resin appeared to give the most consistent applied results. Treatment of the tannin before preparation of the adhesive, although presenting the advantage of eliminating possible internal rearrangements [38,39], also presents the disadvantage that the rearranged tannin will tend to react with formaldehyde through internally generated resorcinol during the subsequent preparation of TRF resin. This partially obviates the advantage of having the internally generated resorcinol available for curing the adhesive [38,39]. Treatment after preparation of the TRF resin has the disadvantage that complications due to structural rearrangements might occur. However, the experience of Hemingway et al. [40] in obtaining pine TRF cold-set formulations by direct grafting of resorcinol to procyanidin tannins by interflavonoid links cleavage indicates that the extent of these rearrangements can be controlled with ease and that their influence on adhesive performance is minimal. Equally, the results obtained on mimosa tannins [38,39] clearly indicate that the influence of such rearrangements can also be controlled with ease. Potentially, there is a difference in the site of linkage of the resorcinol added to the flavonoid (Figs. 7.2 and 7.3). The results obtained with these types of TRF cold-set adhesives were encouraging. In the first case, resorcinol chemical additions as low as 5.5 to 10% based on total liquid resin were achieved [38,39] without apparent loss of performance. This is compared to the present level of 18% resorcinol on total liquid resin at 53% solids, which is now added to commercial TRF cold sets [38,39].

In the second adhesive type (Fig. 7.3), from which externally added resorcinol was completely eliminated [38,39], although the results obtained were lower than those required for cold-set adhesives of structural perfor-mance, the resin did present cold-setting adhesive behavior, indicating the likely liberation of linked resorcinol from the tannin structure. Although short of structural capabilities, the performance of this adhesive [38,39] was not disgraceful and might offer some possibilities for a new generation of tannin-based wood adhesives [38,39]. It must also be pointed out that although

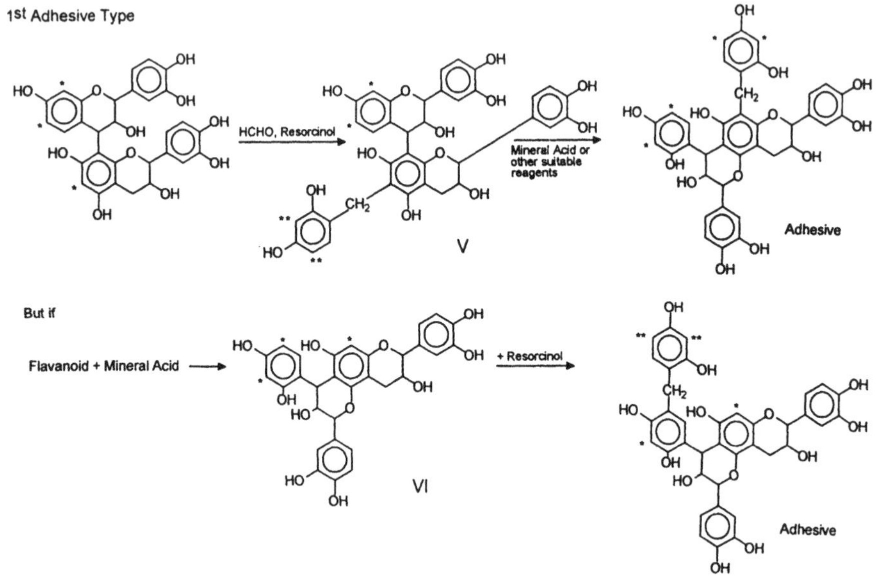

1st Adhesive Type

But if

Flavanoid + Mineral Acid ⟶

Figure 7.2   Two routes that can be used to prepare a TRF cold-set adhesive by use of a photobatannin arrangement.

the results obtained were encouraging, there existed a certain difficulty in obtaining consistently repetitive results with both these systems (i.e., they are more difficult to prepare and use).

V. LRF COLD-SET ADHESIVES

Contrary to the difficulties experienced in the use of lignin as a thermosetting adhesive, lignin–resorcinol–formaldehyde (LRF) cold-set adhesives for application to finger joints and glulam are obtained with ease and work well [41,50]. They satisfy relevant international standard specifications [14,28, 29,41–43]. In particular, lignin-based cold-set and fast-set adhesives in which 13.6% and 19% resorcinol (on total liquid resin) has been grafted onto soda bagasse lignin by first lignin methylolation and then formation of methylene bridges between resorcinol and the methylolated lignin polymer give excellent results. It must be noted that while a traditional-type lignin-based cold set needs only 13.6% resorcinol to compare with synthetic PRF adhesive, the fast-set type needs as much as 19% to obtain an acceptable result [41].

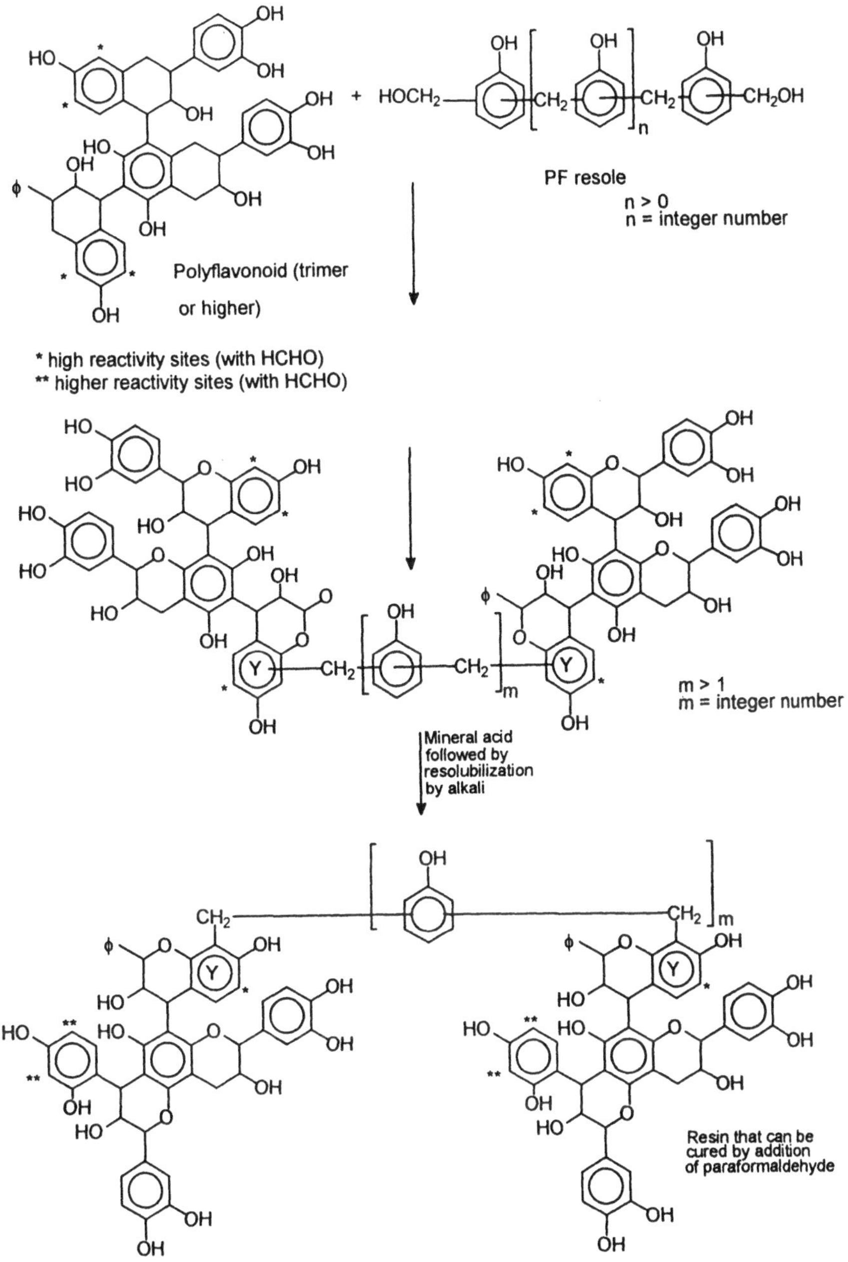

Figure 7.3 Scheme for preparation of mimosa cold-set adhesives by reaction of the tannin with a PF resole followed by acid–base-catalyzed rearrangement to liberate resorcinol functionality.

In these adhesives one resorcinol molecule was introduced for every two phenylpropane units of the soda bagasse lignin. The product is an adhesive that can be cross-linked at ambient temperature by addition of paraformaldehyde, by formation of a series of methylene bridges each connecting two resorcinol terminals each grafted onto a bagasse lignin polymer. The problem of having to operate at a pH somewhat higher (± 9.5) than the pH generally used for PRF adhesives (± 8) was due to the tendency of the lignin to precipitate at lower pHs. To overcome this effect, the adhesive was in a water/methanol 80:20 solution. The amount of methanol present determines the pH at which the traditional-type cold-set lignin adhesive (or component A for the honeymoon types) can be used; when no methanol is present, the pH needed to maintain the polymer in solution is well above 10 and the glue-mix pot life is very short, rendering the adhesive unusable for industrial purposes, notwithstanding the fact that high strengths are still obtained.

As the amount and proportion of methanol is increased, the pH can be lowered considerably, as the presence of methanol allows the lignin polymer to be maintained in solution at much lower pHs. The pot life, at ambient temperature, of the glue mix is extended considerably by both the slowing down of the cross-linking reaction due to the lower pH as well as by the retarding effect that methanol has on this, as well as on all other phenolic cold-set adhesives, due to the formation of unstable hemiacetals [44].

In a 80:20 water/methanol solution, which is quite normal even for PRF adhesives, the glue mix pot life for the lignin-based adhesive is generally ± 88 min at 25°C, which is acceptable for industrial purposes. An increase in the proportion of methanol will lengthen even further the pot life of the glue mix. Attempts were also made to extend the pot life at the higher pH values, where methanol is not required to maintain the polymer in solution, by using several types of slow-release paraformaldehyde [41]. Unfortunately, concomitant to the relative lengthening of the pot life obtained, a decrease in both strength and wood failure of the joints bonded with such glue mixes resulted. Thus slow-release paraformaldehydes do not appear to be usable at this stage to lengthen the pot lives of these adhesives.

In Fig. 7.4 is shown a comparative differential scanning calorimetry examination of a resorcinol/MBL graft adhesive and a traditional PRF cold-set adhesive. Exotherms are present at 90 to 92°C and at 86 to 88°C for the PRF and lignin-based adhesive, respectively. The exotherm is more pronounced for the PRF adhesive than for the lignin-based adhesive: This is consistent with the higher resorcinol percentage on liquid resin of the PRF adhesive (16.5% for PRF against 13.6% for the lignin-based adhesive).

One point that should be noted is that the soda bagasse lignin shows some

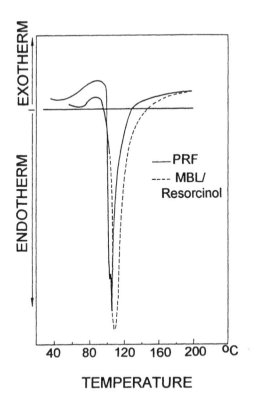

Figure 7.4 Differential scanning calorimetry of lignin-based adhesive and a commercial PRF cold-set adhesive.

sign of cleavage in an alkaline medium at high temperature for a prolonged period. For instance, when methylolation was carried out at reflux over a period of 2 to 3 h, high-pressure gel permeation chromatography clearly indicated that degradation of the lignin occurred—thus the necessity to methylolate under mild conditions to avoid degradation of the polymer. Economically, the resorcinol/MBL graft is particularly attractive.

VI. α-SETTING OF RESORCINOL RESINS

In view of the α- and β-set acceleration of PF resins curing by esters [34], it is interesting to discuss if resorcinol resins are also subject to α-set acceleration by esters. For the reaction of formaldehyde with a resorcinol chemical (not a resin), the α-set acceleration effect has been found to appear only above pH values of 12 to 13 and to be quite small [34]. This is because slowing down the resorcinol reaction with formaldehyde appears only at these very high pHs, and consequently only there can an acceleration effect be noted. The situation is somewhat different for RF and PRF resins. When a preformed resin is involved, the slowing down of the reaction of hardening already appears at pH 10.5 to 11, and consequently, the α-set accelerating effect also becomes evident from this pH range [35]. From this it can be concluded that α-set acceleration in the faster-reacting resorcinol occurs in the same pH ranges as those observed for thermosetting PF resols. However, due to the very fast reactivity of RFs and PRFs, hence of the asymptotic acceleration with increasing pH of these resins to very short gel times, such an effect is not noticeable, and in reality is inconsequential to the rate of curing, until the slowdown in reactivity at very high pH. At these very high pH ranges, as the rate of curing slows, acceleration by α-set mechanisms becomes progressively more noticeable and more significant.

VII. FT-IR and ^{13}C NMR OF RESORCINOL ADHESIVES

In the case of resorcinol resins, PRFs, URFs, and TRFs, it is not possible to obtain from their ^{13}C NMR equations correlating quantitatively the strength of the hardened resin from the liquid spectra, as has been obtained for the thermosetting PF, MF, and UF resins. This is so because the liquid resins themselves are inactive without the addition of paraformaldehyde hardener: in short, they are not resols. For this reason, as for the thermosetting tannin adhesives, it is only possible to deduce qualitative indications with regard to their cold-set characteristics and strength in their hardened state.

A. Phenol–Resorcinol–Formaldehyde

Although ^{13}C NMR is very useful to identify salient characteristics of PRF resins, it is worthwhile to report a FT-IR method that can be used to determine semiquantitatively from the liquid resin if the strength of the cured PRF adhesive will be good or poor in its hardened state [22,23]. The only noticeable difference between the spectra was that between spectra of resins that gave "good" strengths and those that gave "poor" strengths. The difference lay in the intensities of the IR transmission bands at 1017 and 965 cm^{-1}. In strong resins the band at 1017 cm^{-1} is much more intense than that at 965 cm^{-1}, and in weak resins the reverse is observed. The band at 1017 cm^{-1} is due to 1,2-

and 1,4-substituted aromatic rings due to *o*- and *p*-methylolphenolic nuclei, as well as to contributions by the C–O stretch of the methylol groups. The band at 965 cm^{-1} appears primarily to be the C–O stretch of resorcinol (this peak appears very clearly in resorcinol spectra). From this it is understandable that in a strong adhesive there are sufficient methylol groups to react with most of the resorcinol. Thus very little resorcinol is left unreacted, causing the resorcinol band (965 cm^{-1}) to be less intense than the band at 1017 cm^{-1}. In a weak resin, the number of methylol groups available to react with the resorcinol is not sufficient. Some of the resorcinol remains unreacted, and hence the band at 965 cm^{-1} is more intense than that at 1017 cm^{-1}. The excess of unreacted resorcinol appears to make the resin weak. The unreacted resorcinol reacts preferentially with the formaldehyde in the hardener, starving the resin of hardener, causing the resin to be under-cross-linked and slowing the hardening process. The resin is hence weak. Any resin left at room temperature will continue to resinify with time, and hence the ratio of unreacted resorcinol would decrease with time. Consequently, the intensity of the peak at 965 cm^{-1} did decrease, as expected, relative to that at 1017 cm^{-1}.

After standing at room temperature, "strong" resins gave spectra in which the peak at 1017 cm^{-1} seemed to have disappeared almost completely. The resins continue to polymerize at room temperature, and hence the –CH$_2$OH groups gradually reacted to completion. As a result, the intensity of the peak at 1017 cm^{-1} decreases and eventually disappears.

Studies on the effect of 70% methanol: 30% water solution on the phenolic band at 1017 cm^{-1} were carried out to explain further the behavior noted. First, the solvent alone gives a very strong peak at 1025 cm^{-1}. Second, the band at 1017 cm^{-1} increases with increased percentage of solvent present. From this it was concluded that the peak at 1017 cm^{-1} is due to a phenol/CH$_3$OH adduct in the unreacted (no formaldehyde) phenol/methanol mixture and due to phenol/–CH$_2$OH adduct in the PF resin. It was believed, however, that the decrease in intensity of the peak at 1017 cm^{-1} in the aging PRF resins was due primarily to a decrease in the number of unreacted methylol groups. Very little or no loss of solvent from a resin could occur in a closed container, and hence the solvent effect would play a very limited role.

When it was first observed that weak resins had a more intense band at 965 cm^{-1} and strong resins the reverse, it was thought that one could do quantitative studies to determine the amount of unreacted resorcinol in the liquid resin. This, in turn, would help to explain quantitatively why certain resins were weak. Experimental studies [22,23] indicated that the more resorcinol is added to the solution, the greater is the intensity of the peak at

965 cm^{-1} relative to the peak at 1017 cm^{-1}. A graph of the ratio of percentage transmittance of the band at 1017 cm^{-1} against percentage unreacted resorcinol in the liquid resin lead to an exponential regression analysis equation [23]:

$$y = 45.1e^{-0.25x}$$

of coefficient of correlation $r = 0.87$, where y is the ratio of the percent transmittance of the band at 965 cm^{-1} to the percent transmittance of the band at 1017 cm^{-1}, and x is the percent unreacted resorcinol in the liquid resin. This was then related to some strong and weak resins to see if the difference in these resins could be related to the percentage unreacted resorcinol present in the liquid resin. The results obtained indicated that a quantitative correlation is not as simple as originally hoped.

The errors arise due to the simplification that the band at 965 cm^{-1} is due only to the C–O stretch of the unreacted resorcinol. It may, instead, be due to both reacted and unreacted resorcinol, and the degree of absorption of reacted resorcinol may differ from that of unreacted resorcinol. A second error could be introduced by the solvent effect on the band at 1017 cm^{-1}. As discussed earlier, the percentage transmittance of the band at 1017 cm^{-1} is influenced by the amount of solvent present. Experimentally, it is very difficult to obtain a constant percent solids content for all the resins, and thus the amount of solvent in different resins varied. This could influence the percent transmittance of the band at 1017 cm^{-1}, hence introducing an error in the quantitative results obtained above.

However, one cannot say that the ratio of percent transmittance of the band at 965 cm^{-1} to the percent transmittance of the band at 1017 cm^{-1} is totally insignificant. It appears that if this ratio is less than 1, the amount of resorcinol is in excess of the –CH$_2$OH reactive groups, indicating that the resin bond strengths will be poor. However, if the ratio is greater than 1, the number of –CH$_2$OH reactive groups is sufficient to use up the resorcinol added, indicating a resin with good bond strengths. Figure 7.5a reports the CP–MAS ^{13}C NMR spectrum of a hardened PRF resin. In the latter (Fig. 7.5a) it is possible to note ortho–ortho, ortho–para, and para–para linkages, methylene bridges, methylene–ether bridges, and quinonic structures still present in the hardened state.

B. Urea–Resorcinol–Formaldehyde

URF resins are more amenable to analysis of performance by ^{13}C NMR than by IR. For instance, it was not possible to distinguish between mono-, di-, and trisubstitution on the resorcinolic nucleus by change in the chemical shift of the resorcinolic carbon atoms in ^{13}C NMR spectroscopy. Thus ^{13}C NMR

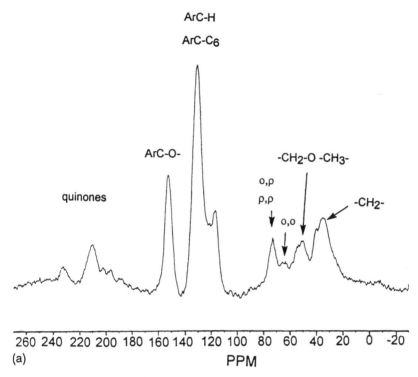

Figure 7.5   (a) CPMAS ^{13}C NMR spectrum of a hardened commercial PRF resin; (b) ^{13}C13 NMR spectrum of a traditional PRF resin in liquid form.

spectroscopy is not used to identify tridimensional branching or linear lengthening on the resorcinolic nuclei, just on the urea molecules. The peak that is used as an identification whether tridimenstional branching occurs is at $\delta \approx 55$ ppm ($-CH_2-N-CH_2-$).

^{13}C NMR spectroscopy also indicates that when tannin extract is used, tridimensional branching on the urea molecule is encouraged even more than when resorcinol is used. The peak at $\delta = 55$ ppm ($-CH_2-N-CH_2-$) is even more intense in these spectra. This further supports the hypothesis that branching on the polyflavonoid tannin occurs more readily due to the better steric availability of the reactive sites. By examining the ^{13}C NMR spectra it is possible to note differences that can be used to understand why the applied results are such. When a low proportion of resorcinol brancher is used, tridimensional branching on the urea molecule appears to occur as the peak

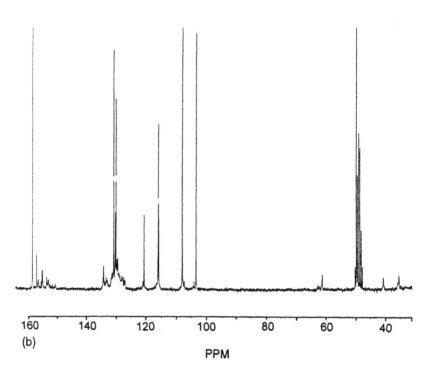

PPM

Figure 7.5   (continued)

at $\delta = 55.90$ to 56.04 ppm is clearly evident. When the amount of resorcinol brancher is further increased, it appears from the ^{13}C NMR spectra that tridimensional branching on the urea molecule is eliminated as the $-N-(CH_2)_2$ peak at $\delta = 56$ ppm disappears. The amount of branching resorcinol may be sufficient to allow the macromolecule to lengthen linearly. This linear lengthening appears to decrease the percent wood failure, probably as a consequence of the decreased ability of the polymer to wet the substrate. Although, as expected, linear lengthening is the favorite mechanism for decreasing the resorcinol required, it appears that in branched URF resins the macromolecules produced are too large. This causes poor specific adhesion and is manifested in poor wood failure results.

^{13}C NMR spectroscopy also indicates a possible explanation for the methanol interference reported. With an increase in the percentage methanol used in the solvent, there is an increase in the amount of monomeric urea and methylenebisurea present. This is deduced by an increase in intensity of the bands at δ = 164.10 ppm (C=O of urea) and δ =161.99 ppm (C=O peak of methylenebisurea). The presence of this increased amount of monomeric urea and methylenebisurea may be causing weakening in the water-resistant properties of the resin.

From the ^{13}C NMR results one can only conclude that too large a percentage of methylenebisurea in the resin appears to have a weakening effect on the water-resistance properties of the resin. This is perhaps the first clear analytical indication of the water lability of the –NH–CH$_2$-NH– bond [23,24]. The relevant parts of a complete ^{13}C NMR spectrum of a URF cold-set adhesive is shown in Figs. 7.6 and 7.7.

C. Tannin–Resorcinol–Formaldehyde

The ^{13}C NMR of a TRF cold-set adhesive is shown in Fig. 7.8. In this spectrum the tannin identification pattern and the resorcinol-based pattern are clearly superimposed [36].

VIII. RFURFURAL AND PRFURFURAL ADHESIVES [45]

RFurfural, RPFurfural, and PRFurfural resins capable of giving good cold-setting exterior-grade adhesives for glulam and finger jointing can be produced as easily as, and with performance equivalent to that of similar resins produced with formaldehyde [45]. One of the difficulties noted is the lower viscosities obtained with the furanic version of these phenolic resins. This minor problem was eliminated by adding small amounts of a postthickener. The addition of the postthickener is necessary to obtain good performance of the resins and to maintain the resin solids content within reasonable and economical limits. The absence of postthickener causes considerable decrease in the strength as well as wood failures that result because the resins, which are too thin, tend to be absorbed by the wood, with consequent glue-line starvation.

The resorcinol content of the total liquid resin is quite low (i.e., 20.2%) and the resins satisfy the exterior-grade structural adhesive requirements of the South African Bureau of Standards SABS 1348-1981 [46] specifications for phenolic resins for wood and of the milder British Standard BS 1204-1965 [47] specifications for synthetic adhesives for wood, in terms of their weather-resistant and boil-proof characteristics. The pot lives obtained are comparable to those obtained with commercial PRF adhesives [45]. The results obtained indicate that PRFurfural resins tend to have lower shrinkage on hardening than that of equivalent PRFormaldehyde resins. This could indicate a better gap-filling capability for these resins. This advantage is

Figure 7.6 ^{13}C NMR spectrum of 44 to 61 ppm region of a URF resin. Note the band at 56.04 ppm, indicating branching.

probably due to the higher solids content afforded by the use of furfural instead of formaldehyde. The higher solid content of these resins does not render them uneconomical. Two counterbalancing effects are present: (1) the percentual decrease of expensive resorcinol in total resin solids due to the higher amount of aldehyde used in the furanic modification (furfural higher

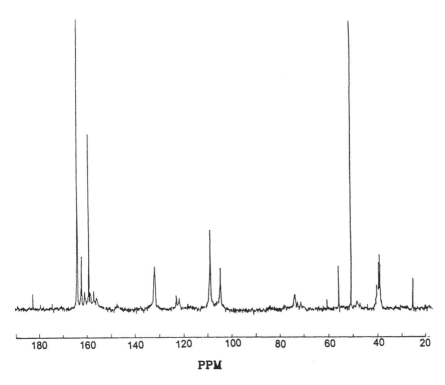

Figure 7.7 ^{13}C NMR spectrum of a tannin-branched URF.

molecular weight), which lowers the resin cost; and (2) the increase in resin solids, which tends to increase the cost. The cost of furanic and formaldehyde cold-setting resins is now practically identical. There are, however, a few advantages in the use of the furanic PRF resins: (1) the lower shrinkage on hardening, which should lead to better gap filling properties; (2) the decreased amount of total formaldehyde in the glue mix; and (3) the use of an aldehyde that can be obtained from waste biomass, such as furfural. The choice of acid rather than alkaline conditions during the first part of the reaction, the phenol–furfural condensation, was adopted to minimize excessive branching of the resin (in acid conditions, mainly linear resins are formed), which may have caused poor tolerance to dry-out in the finished adhesive [48].

The switch to furfural from formaldehyde also appears to be easy in the preparation of TRF resins [49]. Wattle tannin–resorcinol–furfural resin results are comparable to those obtained with TRF and PRF resins. Interesting is the

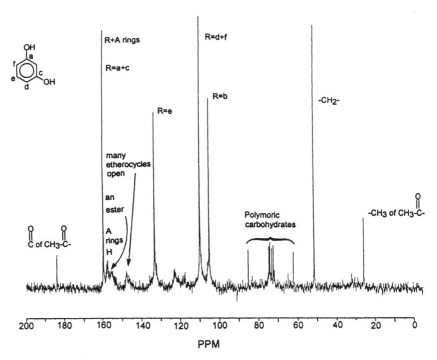

Figure 7.8 ^{13}C NMR spectrum of a tannin–resorcinol–formaldehyde cold-set adhesive.

fact that the tannin–resorcinol–furfural copolymer can also be prepared by acid condensation (it is later adjusted to alkaline pH), not only by alkaline condensation, still obtaining good strength and wood failure results. This may well improve further the dry-out characteristics of tannin-based cold-set resins, as more linear and less branched copolymers are likely to be formed by acid condensation [48].

The ultimate strength of a furfural- rather than a formaldehyde-based honeymoon fast-set system is also excellent. The furfural-based resin appears to be slower setting. This statement needs further clarification: the rate of strength development with time of the furfural-based resin shows a longer induction time before the setting reaction starts: ±30 min for furfural-based resin versus 5 to 10 min for formaldehyde-based resin. This behavior is somewhat surprising if one considers that paraformaldehyde is the hardener used for both resins. It may be interpreted as a function of the bulkier nature of the furfural-based polymer and the lower amount of water present in it, leading, possibly, to slower hydrolysis of the paraformaldehyde hardener.

Comparative analysis by differential thermal analysis of phenol–resorcinol–furfural and of two equivalent commercial PRF resins (Fig. 7.9) shows that while the furfural-based resin presents a fairly broad exothermic peak at 62°C (42 to 87°C) and two endothermic peaks at 103°C (water) and 165°C, the two phenolic resins have similar but shifted patterns [45]. The first commercial PRF resin presents a fairly sharp exothermic peak, of 92°C (77 to 102°C) and the water endotherm at 105°C. The second commercial PRF resin presents a system of two broad combined exothermic peaks at 68°C and 84°C (40 to 98°C for the system), with the 84°C peak being the main one, and two endotherms at 102°C (water) and 193°C, respectively. These data indicate that the furfural-based resin is broadly similar to the two commercial PRF

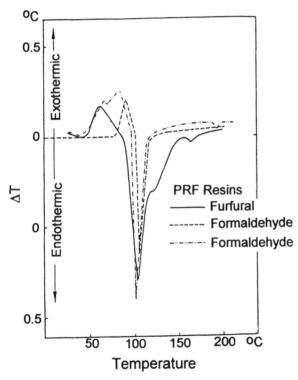

Figure 7.9 Comparative DTA analysis of a phenol–resorcinol–furfural resin and two commercial phenol–resorcinal–formaldehyde resins.

resins. The somewhat lower temperature exotherm for the furfural-based resin appears to indicate a slightly higher amount of reactive material in this resin. As the resorcinol content of the three resins is known to be comparable and the pH levels used during the analysis identical, the difference may be due to the higher solid content of the furanic resin as well as participation of the furanic residues in the cross-linking reaction during setting [45].

To conclude, phenol–resorcinol–furfural resins appear to have the same performance characteristics as those of commercial PRF resins, at the same cost, with a few, perhaps minor advantages. The switch to furfural from formaldehyde appears to present no difficulty in the preparation of tannin–resorcinol cold-setting adhesives as soon as the formaldehyde is substituted by furfural on a molar basis. The use of phenol–resorcinol–furfural resins for application to honeymoon fast-setting adhesives does not appear to present any difficulty, although these resins are slightly slower than PRF fast-set adhesives.

IX. FORMULAS

A. Linear PRF Resin [22]

A basic formulation capable of giving more than adequate results is presented here to acquaint a beginner in the field with these types of adhesives. The procedure for the preparation of this resin can be modified in many ways by varying catalysts, concentration, molar ratios, and condensation conditions.

Phenol, 110 parts by mass + 22 parts water
First formalin 37% solution, 49 parts by mass
H_2SO_4 10% solution, 22 parts by mass
First NaOH 40% solution, 4.5 parts by mass
Second formalin 37% solution, 90 to 93 parts by mass
Methanol or methylated spirits, 30 parts by mass (at start of reaction)
Resorcinol, 71 parts by mass
Tannin extract, 19 parts by mass

Phenol, water, methanol, and the first amount of formalin solution are charged on the reaction vessel and heated mildly until the phenol is dissolved. H_2SO_4 is added and the temperature increased to reflux. The mixture is refluxed for 3.5 to 4 h (generally ±4 h). It is cooled to 50 to 60°C and the the two amounts of NaOH 40% solution (slowly) and the second amount of formalin solution are added. The mixture is refluxed for 4.5 to 4.75 h, and then resorcinol is added. The mixture is refluxed for a further 30 to 50 min.

Spray-dried tannin extract is added immediately before or during cooling to adjust viscosity. The pH must be adjusted to 8.5 to 9.5 according to the pot life required. The hardener is a 50:50 mixture of paraformaldehyde 96% (usually, a fast grade) and 180 to 200 mesh wood flour (60:40 mass proportion) used in 5:1 or 4:1 w/w on liquid resin.

REFERENCES

1. J. M. Dinwoodie, Chapter 1 in *Wood Adhesives: Chemistry and Technology*, Vol. 1 (A. Pizzi, ed.), Marcel Dekker, New York, 1983, pp. 1–58.
2. R. E. Kreibich, in *Wood Adhesives: Present and Future* (A. Pizzi, ed.), Applied Polymer Symposium No. 40, 1984, pp. 1–18.
3. A. Pizzi, D. du T. Rossouw, W. Knuffel, and M. Sigmin, *Holzforsch. Holzverwert.*, *32*(6): 140 (1980).
4. A. Pizzi, R. M. Horak, D. Ferreira, and D. G. Roux, *Cellu. Chem Technol.*, *13*: 753 (1979); *J. Appl. Polym. Sci.*, *24*: 1571 (1979).
5. R. A. V. Raff and B. M. Silverman, *Ind. Eng. Chem.*, *43*: 1423 (1951).
6. D. du T. Rossouw, A. Pizzi, and G. McGillivray, *J. Polym. Sci. Chem. Ed.*, *18*: 3323 (1980).
7. A. Pizzi and P. van der Spuy, *J. Polym. Sci. Chem. Ed.*, *18*: 3477 (1980).
8. R. A. V. Raff and B. H. Silverman, *Can. Chem.*, *29*: 857 (1951).
9. A. R. Ingram, *Can. Chem.*, *29*: 863 (1951).
10. C. T. Liu and T. Naratsuka, *Mozukai Gakkaishi*, *15*: 79 (1969).
11. P. H. Rhodes, *Mod. Plast.*, *24*(12): 145 (1947).
12. R. H. Moult, Chapter 24 in *Handbook of Adhesives*, 2nd ed. (I. Skeist, ed.), Van Nostrand Reinhold, New York, 1977, pp. 417–423.
13. G. G. Marra, *For. Prod. J.*, *6*: 97 (1956).
14. A. Pizzi, Chapter 3 in *Wood Adhesives: Chemistry and Technology*, Vol. 1 (A. Pizzi, ed.), Marcel Dekker, New York, 1983, pp. 105–178.
15. G. F. Baxter and R. E. Kreibich, *For. Prod. J.*, *23*(1): 17 (1973).
16. R. W. Caster, *For. Prod. J.*, *23*(1): 26 (1973).
17. H. Ericson, *Papper Och Tra*, *1*: 19 (1975).
18. R. E. Kreibich, *Adhes. Age*, *17*: 26 (1974).
19. A. Pizzi and F. A. Cameron, *For. Prod. J.*, *34*(9): 61 (1984).
20. A. Pizzi and F. A. Cameron, Chapter 9 in *Wood Adhesives Chemistry and Technology*, Vol. 2 (A. Pizzi, ed.), Marcel Dekker, New York, 1989, pp. 229–306.
21. A. Pizzi, Chapter 7 in *Wood Adhesives Chemistry and Technology*, Vol. 2 (A. Pizzi, ed.), Marcel Dekker, New York, 1989, pp. 190–210.
22. E. Scopelitis and A. Pizzi, *J. Appl. Polym. Sci.*, *47*: 351 (1993).
23. E. Scopelitis, M.Sc. thesis, University of the Witwatersrand, Johannesburg, South Africa, 1992.
24. E. Scopelitis and A. Pizzi, *J. Appl. Polym. Sci.*, *48*: 2135 (1993).

25. E. A. Blommers and R. H. Moult, U.S. patent 4,032,515 (1977).
26. J. R. Ebdon and P. E. Heaton, *Polymer*, *18*: 971 (1977).
27. R. M. Rammon, W. E. Johns, J. Magnuson, and A. K. Dunker, *J. Adhes.*, *19*: 115 (1986).
28. *Specification for Synthetic Resin Adhesives*, British Standard BS 1204-1965, Parts 1 and 2, 1965.
29. *Phenolic and Aminoplastic Adhesives for Laminating and Fingerjointing of Timber, and for Furniture and Joinery*, South African Bureau of Standards, SABS 1349-1981, 1981.
30. A. Pizzi, Chapter 4 in *Wood Adhesives: Chemistry and Technology*, Vol. 1 (A. Pizzi, ed.), Marcel Dekker, New York, 1983.
31. N. J. L. Megson, *Phenolic Resin Chemistry*, Butterworth, London, 1958.
32. A. Pizzi, *J. Adhes. Sci. Technol.*, *4*: 573 (1990); *4*: 589 (1990); *Holforsch. Holzverwert.*, *43*(3): 63 (1991).
33. M. Saula, Chapter 25 in *Handbook of Adhesives*, 2nd ed. (I. Skeist, ed.), Van Nostrand Reinhold, New York, 1977.
34. A. Pizzi and A. Stephanou, *J. Appl. Polym. Sci.*, *49*: 2157 (1993); *Holzforschung 48*: 150 (1994).
35. A. Pizzi and A. Stephanou, unpublished work, 1993.
36. D. Pienaar, unpublished results, 1992.
37. D. G. Roux, in *Adhesives from Renewable Resources* (R. W. Hemingway and A. H. Conner, eds.), ACS Symposium Series No. 385, American Chemical Society, Washington, D.C., 1989.
38. A. Pizzi, F. A. Cameron, and E. Orovan, *Holz Roh Werkst.*, *46*: 67 (1988).
39. A. Pizzi, in *Plant Polyphenols* (R. W. Hemingway and P. E. Laks, eds.), Plenum Press, New York, 1992.
40. R. E. Kreibich and R. W. Hemingway, *For. Prod. J.*, *35*(3): 23 (1985); *43* (1993).
41. G. H. van der Klashorst, F. A. Cameron, and A. Pizzi, *Holz Roh Werkst.*, *43*: 477 (1985).
42. *Code of Practice for the Terminology and Classification of Adhesives for Wood: Specification for Exterior/Structural Adhesives for Wood*, SABS 0183-1981, South African Bureau of Standards, 1981.
43. *Code of Practice for the Manufacture of Fingerjointed Structural Timber*, SABS 096-1976, South African Bureau of Standards, 1976.
44. J. F. Walker, *Formaldehyde*, 3rd ed., R. E. Kreiger, New York, 1975.
45. A. Pizzi, E. Orovan, and F. A. Cameron, *Holz Roh Werkst.*, *42*: 467 (1984).
46. *Standard Specification for Phenolic and Aminoplastic Resin Adhesives for the Laminating and Fingerjointing of Timber, and for Furniture and Joinery*, SABS 1348-1981, South African Bureau of Standards, 1981.
47. *Standard Specification for Synthetic Resin Adhesives for Wood*, British Standard BS 1204-1965, 1965.
48. L. Gollob, Ph.D. thesis, Oregon State University, Corvallis, Oreg., 1983.
49. A. Pizzi and D. G. Roux, *J. Appl. Polym. Sci.*, *22*: 1945 (1978).
50. P. Truter, A. Pizzi, and H. Vermaas, *J. Appl. Polym. Sci.*, *51*: 1319 (1994).

8

Diisocyanate Wood Adhesives

I. INTRODUCTION

The use of diisocyanates for wood adhesives is relatively recent [1–3], the commercial manufacture of diisocyanate-bonded particleboard having started in Germany in 1975. For several years while considerable research interest was focused on their use as wood adhesives, the literature on the subject being quite vast, these materials struggled to assert themselves industrially as the excellently performing adhesives they really are. Thus industrial use was slow to expand, although today, diisocyanates are used quite widely and extensively to produce exterior-grade particleboard. This initial, possibly unwarranted struggle to be widely accepted as thermosetting wood adhesives has been due to several factors: (1) their initial characteristic of tending to bind the board to the press, requiring that the board surfaces be glued with a different type of adhesive; (2) their toxicity and low vapor pressure; (3) the impossibility of using them in plywoods; (4) the impossibility of diluting them with water, a problem partially solved by the introduction of emulsified diisocyanates [3]; (5) their higher cost; and (6) the normal resistance to the introduction of such a different bonding system. Some of these problems have been solved, and solved well, as the later expansion of diisocyanates used in the wood industry clearly demonstrates; others persist. There is no doubt that these are excellent wood adhesives: it was assumed that such strong bonding could only be obtained by the formation of a high proportion of covalent bonds between the resin and the lignocellulosic substrate [1–4]. This assumption is particularly seducing, as wood is literally teeming with the types of chemical groups, hydroxy groups, with which the isocyanate favors reaction. However, this is just an assumption as there is no experimental evidence of real relevance for it to their use as thermosetting wood adhesives for particleboard.

II. CHEMISTRY OF DIISOCYANATE ADHESIVES BONDING

The isocyanate group reacts readily with any hydroxy group,

$$R–N=C=O + HO–R' \rightarrow R–NH–\overset{\overset{\displaystyle O}{\|}}{C}–O–R' \qquad (8.1)$$
$$\text{urethane}$$

to form a urethane bridge. The readiness of reaction between isocyanate and hydroxyls extends to water, with which the isocyanate group reacts readily, with the liberation of carbon dioxide and simultaneous formation of substituted urea groups. The first step in this reaction is the formation of unstable carbamic acid, which decomposes to form an amine and carbon dioxide.

$$R–N=C=O + H_2O \rightarrow [R–NH–\overset{\overset{\displaystyle O}{\|}}{C}–OH] \rightarrow R–NH_2 + CO_2 \qquad (8.2)$$

As primary and secondary amines are the other favorite group with which isocyanates react, the amine formed by the reaction above reacts immediately with additional isocyanate to form a substituted urea as follows:

$$R–NH_2 + O=C=N–R \overset{\text{fast}}{\longrightarrow} R–NH–\overset{\overset{\displaystyle O}{\|}}{C}–NH–R \qquad (8.3)$$
$$\text{substituted urea}$$

The secondary amine groups present in the urethane bridge [reaction (8.1)] and in the substituted urea [reaction (8.3)] formed react further with available isocyanate groups to continue cross-linking and hardening of the material by the formation of allophenate and biuret bridges:

$$R–N=C=O + R–NH–CO–OR' \rightarrow R–NH–CO–\underset{\underset{\displaystyle R}{|}}{N}–CO–OR' \qquad (8.4)$$
$$\text{allophenate}$$

$$R-N=C=O + R-NH-CO-NH-R \rightarrow R-NH-CO-N-CO-NH-R \qquad (8.5)$$
$$\underset{\text{biuret}}{\overset{|}{R}}$$

The progress of reactions (8.2) to (8.5) is indicative that even traces of water can set off a chain of events in which the hydroxy groups of wood do not need to participate at all in the cross-linking of the resin. If one observes (1) the hardness of a diisocyanate gel to which only drops of water have been added, and if it is considered that (2) reaction of the isocyanate group with primary and secondary amines is markedly faster than with hydroxyls, the assumption of a preponderance of covalent bonds between resin and lignocellulosic substrate no longer appears a serious possibility, especially at the very short pressing times characteristic of particleboard. Traces of water always present in wood, and even the formation of a few initial covalent bonds between the still-liquid resin and the lignocellulosic substrate, will then set off a chain of reactions in which the material needs only the amino groups formed by its own molecules to harden. Thus

$$R-NCO + H_2O \rightarrow R-NH_2 \xrightarrow{+R-NCO} R-NHCONH-R \xrightarrow{+R-NCO} RNHCONCONHR$$

$$\underset{R}{\overset{|}{\downarrow}} {+R-NCO}$$

$$R-NHCO-[NCO]_n-NHR \xleftarrow{+_nR-NCO} R-NCONCONCONHR$$
$$\underset{R}{\overset{|}{}} \qquad \underset{R\ R}{\overset{|\ \ |}{}}$$

As to each R belong at least two –NCO groups, a hard, cross-linked network is formed rapidly. Equally, starting from reaction with an hydroxyl,

$$R-NCO + HO-R' \rightarrow R-NHCOOR' \rightarrow R-NHCON-COOR'$$
$$\underset{R}{\overset{|}{}}$$
$$R-NHCO-[NCO]_n-OR'$$

Thus just one molecule of water, or reaction with a hydroxyl molecule of wood would be more than enough to gel and harden the material without imagining any covalent bonding between resin and substrate except perhaps one (in a multitude of other resin-to-resin bonds), the first.

There are two reasons why such a possibility has not been considered seriously before:

1. For many years the misconception has existed that even formaldehyde-based resins form covalent bonds between resin and substrate in order to

provide a "good bond." This is incorrect, particularly so in the very rapid curing environments prevalent for thermosetting wood adhesives. Past work "proving" covalent bond formation had to go to the ridiculously long curing times of 2 to 12 h at 120°C to find covalent bonding [5–7]: in industrial particleboard, the board core, which determines the board's internal bond, achieves 105°C, possibly 110°C, for less than 30 to 60 s! Even after 2 h at 120°C it has been clearly demonstrated that the adhesive forms a covalent bond between resin and wood every 1000 to 1600 resin-to-resin bonds [8]. More recent work has shown the covalent bond proportion to be from very, very small to nonexistent [9–11].

2. The word *urea* (or *polyurea*) conjures up for wood adhesives technologists the idea of non-water-resistant adhesives, from their knowledge of UF resins. In UF resins the liability to water attack is due exclusively to the aminomethylenic bond, thus to the susceptibility to water of the $-H_2C-NH$ bond. This bond does not exist in polyureas derived from diisocyanate reaction with water or hydroxyl groups. After all, in urea itself, the H_2N-CO bond is not hydrolyzable under any circumstances: thus true polyureas are not hydrolyzable, offering very strong adhesion to the substrate. Otherwise, how could an isocyanate adhesive be able to give good strong bonds between the metal of a hot-press platen and the board surface, considering that there are no hydroxy groups on the metal surface? There are, of course, other reactions contributing to cross-linking to strong networks of diisocyanate adhesives, such as trimerization to isocyanurate rings [4], which are hydrolytically and thermally very stable.

All the above does not exclude the possibility of a certain proportion of covalent urethane bridges forming between the diisocyanate adhesive and the hydroxyls of the lignocellulosic substrate: it serves only to put into perspective under which reaction conditions a proportion of covalent resin/wood bonds might or might not be formed. A recent DSC study has addressed this, particularly for isocyanate adhesives [12]. The study found that a certain, minor proportion of covalent bonds between isocyanate resin and wood substrate are indeed formed. It also found that the major contribution to the strength of isocyanate wood adhesives is the catalytic effect exercised by wood, particularly by wood carbohydrates, in promoting autocondensation of isocyanates to polyureas according to what is presented in reactions (8.2) to (8.5) [9,12]. It is clear, then, that application conditions are determining as regards the presence or absence of resin to substrate covalent bonding. For example, the surface bonding of particleboard in direct contact with a press platen at 200°C for several minutes will definitely present a proportion of covalent bonding between resin and wood substrates [12]. In the core bonding of a particleboard, at 105°C for 30 to 60 s, covalent bonds will definitely be absent or present only in traces [12]. The foregoing perspective is necessary to redress prevalent and current misconceptions in the field of wood bonding with diisocyanates.

III. DIISOCYANATE COPOLYMERIZED WITH PFs, UFs, TFs, LIGNIN–FORMALDEHYDE, AND STARCHES

A growing body of literature exists in the use, and copolymerization, of polymeric diisocyanates (MDIs) with a variety of other resins to yield thermosetting wood adhesives of excellent performance. The reasons for the growing interest in this type of copolymers are several and very relevant: (1) to prepare adhesives of lower cost than polymeric MDI but which equal its excellent performance (i.e., diisocyanates are expensive); (2) to decrease by copolymerization the inherent toxicity of MDI; (3) to allow the use of polymeric MDI for plywood, where it cannot be used by itself, due to its low viscosity, high mobility, and consequent very poor performance, hence as an upgrading and accelerating agent of more traditional plywood resins on veneer species that are difficult to bond; and (4) to upgrade the performance or ease of handling of cheaper copolymers, such as natural origin materials to a better level than corresponding synthetic resins.

Several of these uses have already been described in some detail in preceding chapters in this volume. Here an overview is presented to give a nonfragmentary look at this field. Up to very recent times, knowledge of diisocyanates and polyurethane chemistry led to the assumption that in mixed resins containing MDI–PF, MDI–MF, or MDI–UF resins, the isocyanate group could only react with water, leading to MDI deactivation [15]. What had always been disregarded was the fact that the isocyanate group can and does react extremely rapidly with the methylol group ($- CH_2OH$) of PF resols and with the equally reactive methylol group present in MF and UF resins [15–17].

This finding, first observed in 1981 [13,14], and the reactions on which it is based [15], opened the door to a considerable amount of experimentation. The first of these adhesives to have come to industrial fruition was the MDI–PF system for plywood [16] and the MDI–TF system for particleboard [17], both in 1990. In the first of these the methylol-carrying PF resol undergoes preferential reaction with the isocyanate group at the hydroxybenzyl alcohol moieties of the PF resin: this reaction is faster and more favorable than the otherwise expected deactivation of MDI in water. For the first time, this copolymer made possible the industrial use of MDI for plywood, and its industrial conditions of application and performance are now well documented [15,16]. It allowed bonding to marine-grade standards of veneer species for which even the best PF resin could not give satisfactory results, at faster pressing times, higher moisture content of the veneer, and by using much simpler and cheaper PF resins [16].

The second system, the MDI/tannin–formaldehyde combination, allowed the first industrial use of pine tannins as adhesives for exterior particleboard to an excellent level of performance [17]. Notwithstanding the existence today of better and cheaper pine tannin adhesive formulations in which diisocyanates are not used [18,19], this formulation is still used industrially in South America, where it had its first commercial use [16,17].

Other methylol-carrying resins have proved excellent in combination with small amounts of diisocyanates. The most interesting recent application is for upgrading the methylolated lignins to full standard performance [20]. Methylolated kraft lignins [20,21], bagasse lignin [20,21], and even organosolv lignin [22] performed as well as sophisticated, expensive synthetic PF resins once coreacted with small amounts of polymeric MDI. Adhesives for particleboard of excellent exterior performance and fast pressing times in which the lignin content was higher than 60% were prepared with this system [20,21].

Use of MDI to upgrade UF resins to melamine, or better, MUF performance for plywood applications have also been reported [15]. MDI has also been used to upgrade starch for application to plywood [23], with interesting results. The chemistry of this reaction has not been clarified, and it is not known if the isocyanate groups react preferentially with the numerous nonaromatic – –CH_2OH groups or with any –OH present on the carbohydrates. Although this question is purely of academic interest, it would be worthwhile to know if nonaromatic, nonamidic, and nonaminic hydroxymethyl groups are also responsible for fast reactions with –NCO groups to counteract water deactivation of MDI.

IV. CATALYSTS OF THE ISOCYANATE–METHYLOL REACTION

Although MDI–PF adhesive systems are faster curing than PF resins, a wide variety of characteristics can be introduced by the use of PFs presenting different chemical and physical properties. While exterior accelerating agents such as α-set esters [21] are beneficial in the case of slow-reacting PF resins in MDI–PF systems, the amount of accelerator used is generally high, between 5 and 15% of total resin solids. In MDI–PF systems, however, another type of accelerator can be used. The reaction between an isocyanate group and phenolic methylol is nothing but a case of urethane bridge formation, although in water it is possible to use urethane formation catalysts which are routinely used in non-water-based systems [4]. The amount of catalyst used is small, between 0.5 and 1% of total resin solids, and they are of relatively low cost. Such catalysts were also shown to work for MDI–PF adhesive systems in water and were applied with a certain measure of success to exterior wood adhesives for particleboard [24]. In MDI–PF systems using a slow-reacting simple PF resin, their use yielded board pressing times at 200°C as low as 5 s mm^{-1}[24], pressing times faster than those for UF resins. If it is considered that such fast press times were achieved at a 13 to 15% moisture content of the resinated chips, while the boards did not steam blow but rather maintained full exterior-grade properties [24], it is clear that such a catalyst system is superior to the α-set type of acceleration for MDI/synthetic PFs [21]. Such a system is not quite as useful for MDI/methylolated lignin systems, however, because although cure of the resin is definitely accelerated, the catalyst system does not contribute to an increase in the density of cross-linking in the hardened glue line [24] as the α-set system does [20,21]. Such catalysis, then, remains just a manner to even further upgrade the performance of simpler, low-cost synthetic PF resins, its effectiveness decreasing the faster curing and the more sophisticated the PF resin used is [25]. The catalysts that were tried for this system are dibutyltin dilaureate and triethylamine [24]. Faster catalysts [4] could also be used but those might reduce the pot life of the glue mix to unmanageably short periods.

V. BLOCKED MDI–PF THERMOSETTING ADHESIVES

The MDI–PF system for plywood or particleboard (the use of small amounts of tannins are not needed if a fast, sophisticated PF is used [25]) has an inherent advantage that is also, under certain conditions, a limiting disadvantage. While a plywood mill needs only to add raw polymeric MDI to its standard PF glue mix, the inclusion of MDI limits the pot life of the glue mix to approximately 5 to 6 h. This is not problematic for any plywood mill but is problematic for an adhesive manufacturer that would prefer to supply a premixed, stable MDI–PF. This problem can be overcome with ease by the use of phenol-blocked isocyanates or phenol-blocked urethanes [26]. The –NCO group is prereacted with phenol in the absence of water, forming a urethane bridge involving the isocyanate group and the phenolic hydroxyl of phenol itself. Such phenolic-blocked polyurethane is then mixed with the alkaline PF resol in water, and the resulting mixture is almost indefinitely stable, as the isocyanate is "blocked" (protected) against reaction with both water and the methylol groups of the phenolic resin. When the adhesive is used, a small amount of a deblocking agent is used, generally a diamine or *n*-tributylamine, the latter acting both as a catalyst for the reaction of –NCO with the phenolic –OH and with the hydroxybenzyl alcohol group, and as a deblocking agent [27]. This is a very promising system but suffers from the drawback of being quite costly, as blocked diisocyanates are expensive.

~~~~~~~~NCO + HO—⟨O⟩——(anhydrous)→ ~~~~~~~~ NHCOO —⟨O⟩

  blocked isocyanate

~~~~~~~~NCOO—⟨O⟩——(deblocking catalyst)→ ~~~~~ NCO   +   phenol

                                                        OH
                                                         |
  + PF = HOCH$_2$ —⟨O⟩—CH$_2$

                                                        OH
                                                         |
  ~~~~~NHCOO–CH$_2$ —⟨O⟩—CH$_2$~~~~~

This system also suffers from the disadvantage that considerable amounts of free phenol remain in the cured adhesive: a serious problem in view of environmental regulations in many countries.

VI. FURANIC DIISOCYANATES

Furfural is a renewable chemical that is presently produced in modest quantities on a worldwide basis. By-product streams in the forest products industry and other agricultural residues constitute large, untapped reservoirs of feedstock for furfural production. A number of adhesive resins based on furfural and its hydrogenation product, furfuryl alcohol, have been reported in the literature. These resins are utilized primarily as molding resins in the foundry industry but generally have not been acceptable as wood adhesives. Derivatives of furfural, in particular those based on furfurylamine, are possible precursors of resins that might have utility as wood adhesives. Recently, the synthesis of difurfuryl diisocyanates from furfurylamine was reported in the literature. These compounds have reactivities intermediate between those of aliphatic and aromatic diisocyanates. Preliminary experiments strongly suggest that they can be used as wood adhesives [28], although their cost is probably high. Wood adhesives based on reaction of furfuryl alcohol with polymeric diisocyanates have also been reported [29]: these appeared to function well at their experimental stage but do not appear to have been commercialized up to now.

VII. FORMULA

A. MDI–PF Glue Mix for Plywood [15,16,25]

To a fast commercial PF resin for plywood (gel time 24 min at 100°C), 100 parts 45% solids content, are added 11 to 19 parts of raw polymeric MDI, 6.5 to 20 parts by mass of wood flour or coconut shell flour 200 mesh filler, and water to obtain the desired viscosity. This glue mix is applied as such to the veneer plys. A reaction between PF and MDI is visually evident immediately by the formation of a resin of different color. With some PF resins, this might have soft chewing gum–like appearance and viscosity. The addition of small amounts of water and vigorous stirring returns the resin to normal.

REFERENCES

1. H. Deppe and K. Ernst, *Holz Roh Werkst.*, *29*: 45 (1972).
2. H. J. Deppe, *Holz Roh Werkst.*, *35*: 295 (1977).
3. G. W. Ball and R. P. Redman, *Proceedings of the FESYP International Particleboard Symposium*, Hamburg, Sept. 18–20, 1978, pp. 121–126.
4. K. C. Frisch, L. P. Rumao, and A. Pizzi, Chapter 6 in *Wood Adhesives*: *Chemistry and Technology*, Vol. 1 (A. Pizzi, ed.), Marcel Dekker, New York, 1983.
5. S. Chow, *Wood Sci.*, *1*(4): 215 (1969).
6. S. Chow and H. N. Murai, *Wood Sci.*, *4*(4): 202 (1972).
7. A. Gandini, F. Mora, and F. Pla, *Angew. Makromol. Chem.*, *173*: 137 (1989).
8. G. C. Allan and A. N. Neogi, *J. Adhes.*, *3*: 13 (1971).
9. A. Pizzi, B. Mtsweni, and W. Parsons, *J. Appl. Polym. Sci.*, in press (1993).
10. H. Poblete and E. Roffael, *Holz Roh Werkst.*, *43*: 57 (1985).
11. G. E. Meyers, *Wood Adhesives in 1985*: *Status and Needs*, Forest Products Research Society, Madison, Wis., 1986.
12. A. Pizzi and N. Owens, *Holzforschung*, in press (1994).
13. A. Pizzi, *J. Macromol. Sci. Chem. Ed.*, *A16*(7): 1243 (1981).
14. A. Pizzi, *Holz Roh Werkst.*, *40*: 293 (1982).

15. A. Pizzi and T. Walton, *Holzforschung*, *46*: 541 (1992).
16. A. Pizzi, J. Valenzuela, and C. Westermeyer, *Holzforschung*, *47*: 68 (1993).
17. A. Pizzi, E. P. von Leyser, J. Valenzuela, and J. G. Clark, *Holzforschung*, *47*: 168 (1993).
18. A. Pizzi, J. Valenzuela, and C. Westermeyer, *Holz Roh Werkst.*, in press (1993).
19. A. Pizzi and A. Stephanou, *Holz Roh Werkst.*, in press (1994).
20. A. Stephanou and A. Pizzi, *Holzforschung*, *47*: 439 (1993).
21. A. Stephanou and A. Pizzi, *Holzforschung, 47:* 501 (1993).
22. A. Pizzi, J. Valenzuela, J. Baeza, and J. Freer, unpublished results, 1993.
23. R. Marutsky and B. Dix, Chapter 17 in *Adhesives from Renewable Resources* (R. W. Hemingway and A. H. Conner, eds.) ACS Symposium Series No. 385, American Chemical Society, Washington, D.C., 1989.
24. D. B. Batubenga, A. Pizzi, A. Stephanou, P. Cheesman, and R. Krause, *Holzforschung*, in press (1994).
25. B. Dombo, Bakelite A. G., unpublished results, 1993.
26. K. H. Hentschell, E. Jürgens, and W. Wellner, *Farben Lacke*, *2*: 97 (1988); C. R. Vincentz-Verlag, Hanover, *54th Conference of the GFCH Working Group on Coatings and Pigments*, Bad Kissingen, Germany, Sept. 8–10, 1987.
27. J. M. Zhuang and P. R. Steiner, *Holzforschung*, *47* (1993).
28. A. H. Conner, L. F. Lorenz, M. Holfinger, and C. G. Hill, Jr., *19th IUFRO World Congress*, Montreal, 1990.
29. W. McKillip and W. A. Johns, unpublished results, 1980.

Index

283

Milton Keynes UK
Ingram Content Group UK Ltd.
UKHW040447071024
449327UK00020B/1065